Intermediate 2

Chemistry

2001 Exam

2002 Exam

2003 Exam

2004 Exam

2005 Exam

Leckie ✕ Leckie

First exam published in 2001.
Published by Leckie & Leckie, 8 Whitehill Terrace, St. Andrews, Scotland KY16 8RN tel: 01334 475656 fax: 01334 477392
enquiries@leckieandleckie.co.uk www.leckieandleckie.co.uk

ISBN 1-84372-325-5

A CIP Catalogue record for this book is available from the British Library.

Printed in Scotland by Scotprint.

Leckie & Leckie is a division of Granada Learning Limited, part of ITV plc.

Acknowledgements

Leckie & Leckie is grateful to the copyright holders, as credited at the back of the book, for permission to use their material.
Every effort has been made to trace the copyright holders and to obtain their permission for the use of copyright material.
Leckie & Leckie will gladly receive information enabling them to rectify any error or omission in subsequent editions.

[BLANK PAGE]

FOR OFFICIAL USE

Section B | Total Marks

X012/201

NATIONAL QUALIFICATIONS 2001	THURSDAY, 24 MAY 9.00 AM – 11.00 AM

CHEMISTRY
INTERMEDIATE 2

Fill in these boxes and read what is printed below.

Full name of centre

Town

Forename(s)

Surname

Date of birth
Day Month Year

Scottish candidate number

Number of seat

Necessary data will be found in the Chemistry Data Booklet for Standard Grade and Intermediate 2 (1999 Edition).

Section A —Part 1 Questions 1 to 25 and Part 2 Questions 26, 27 and 28

Instructions for completion of **Part 1** and **Part 2** are given on pages two and eight respectively.

Section B (Questions 1 to 16)

All questions should be attempted.

The questions may be answered in any order but all answers are to be written in the spaces provided in this answer book, and must be written clearly and legibly in ink.

Rough work, if any should be necessary, as well as the fair copy, is to be written in this book.

Rough work should be scored through when the fair copy has been written.

Additional space for answers and rough work will be found at the end of the book. If further space is required, supplementary sheets may be obtained from the invigilator and should be inserted inside the **front** cover of this book.

Before leaving the examination room you must give this book to the invigilator. If you do not, you may lose all the marks for this paper.

SCOTTISH
QUALIFICATIONS
AUTHORITY

SECTION A

PART 1

Check that the answer sheet provided is for Chemistry Intermediate 2 (Section A).

Fill in the details required on the answer sheet.

In questions 1 to 25 of this part of the paper, an answer is given by indicating the choice A, B, C or D by a stroke made in INK in the appropriate place in Part 1 of the answer sheet—see the sample question below.

For each question there is only ONE correct answer.

Rough working, if required, should be done only on this question paper, or on the rough working sheet provided—**not** on the answer sheet.

At the end of the examination the answer sheet for Section A **must** be placed **inside** this answer book.

This part of the paper is worth 25 marks.

SAMPLE QUESTION

To show that the ink in a ball-pen consists of a mixture of dyes, the method of separation would be

        A fractional distillation

        B chromatography

        C fractional crystallisation

        D filtration.

The correct answer is B—chromatography. A **heavy** vertical line should be drawn joining the two dots in the appropriate box in the column headed **B** as shown **in the example on the answer sheet.**

If, after you have recorded your answer, you decide that you have made an error and wish to make a change, you should cancel the original answer and put a vertical stroke in the box you now consider to be correct. Thus, if you want to change an answer **D** to an answer **B**, your answer sheet would look like this:

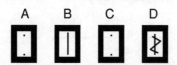

If you want to change back to an answer which has already been scored out, you should **enter a tick (✓)** to the RIGHT of the box of your choice, thus:

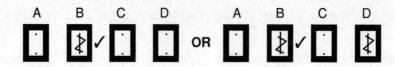

SECTION A

PART 1

1. Which element is an alkali metal?

   A   Aluminium

   B   Calcium

   C   Copper

   D   Sodium

2. Which of the following is the electron arrangement for a metal?

   A   2, 8, 1

   B   2, 8, 5

   C   2, 8, 7

   D   2, 8, 8

3. Metallic bonding is a force of attraction between

   A   positive ions and delocalised electrons

   B   negative ions and delocalised electrons

   C   negative ions and positive ions

   D   a shared pair of electrons and two nuclei.

4. Which of the following groups can react together to form an amide (peptide) link?

   A $-O-H$ and $\overset{H}{\underset{H}{\diagdown}} N-$

   B $-O-H$ and $\overset{O}{\underset{H-O}{\diagdown}} C-$

   C $-N\overset{H}{\underset{H}{\diagup}}$ and $\overset{O}{\underset{H-O}{\diagdown}} C-$

   D $-N\overset{H}{\underset{H}{\diagup}}$ and $\overset{H}{\underset{H}{\diagdown}} N-$

5. Which pair of reactants would produce hydrogen most slowly?

   A   Magnesium powder and 4 $mol\,l^{-1}$ acid

   B   Magnesium ribbon and 2 $mol\,l^{-1}$ acid

   C   Magnesium powder and 2 $mol\,l^{-1}$ acid

   D   Magnesium ribbon and 4 $mol\,l^{-1}$ acid

6. The charge on the iron ion in $Fe_2(SO_4)_3$ is

   A 2^+

   B 3^+

   C 2^-

   D 3^-.

7. In a reaction, 60 cm^3 of gas were collected in 20 s.

   The average rate at which gas was given off, in $cm^3\,s^{-1}$, was

   A $\dfrac{1}{20}$

   B $\dfrac{1}{60}$

   C $\dfrac{20}{60}$

   D $\dfrac{60}{20}$

8. The gram formula mass of sodium carbonate is 106 g.

   How many moles are present in 5·3 g of sodium carbonate?

   A   0·05

   B   0·5

   C   2

   D   20

[Turn over

Questions 9 and 10 refer to the following information.

The fractional distillation of crude oil produces a number of different fractions.

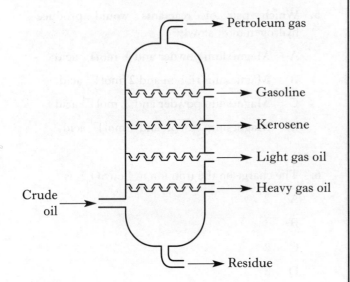

9. Which properties apply to a fraction which has a high boiling point?

    A   High viscosity and low flammability

    B   Low viscosity and low flammability

    C   High viscosity and high flammability

    D   Low viscosity and high flammability

10. Which molecule is most likely to be present in kerosene?

    A C_5H_{12}

    B $C_{12}H_{26}$

    C $C_{19}H_{40}$

    D $C_{26}H_{54}$

11.

Which compound is an isomer of the one shown above?

A

B

C

D

12. A compound burns in air. The only products of the reaction are carbon dioxide, sulphur dioxide and water vapour.

The compound **must** contain

    A   carbon and hydrogen only

    B   carbon and sulphur only

    C   carbon, hydrogen and sulphur

    D   carbon, hydrogen, sulphur and oxygen.

13. One way in which ethanol is produced industrially is shown below.

$$
\begin{array}{c}
\text{H}\quad\text{H} \\
|\quad\ | \\
\text{C}=\text{C} \\
|\quad\ | \\
\text{H}\quad\text{H}
\end{array}
\ +\ \text{H}_2\text{O}
\ \xrightarrow{\text{catalyst}}\
\begin{array}{c}
\text{H}\quad\text{H} \\
|\quad\ | \\
\text{H}-\text{C}-\text{C}-\text{OH} \\
|\quad\ | \\
\text{H}\quad\text{H}
\end{array}
$$

What name is given to this type of reaction?

    A   Condensation

    B   Hydration

    C   Hydrolysis

    D   Oxidation

14. Biopol is a polymer which is

    A   natural and biodegradeable

    B   synthetic and biodegradeable

    C   natural and non-biodegradeable

    D   synthetic and non-biodegradeable.

15. A section of a polymer is shown below.

$$
-\text{O}-\overset{\displaystyle\text{O}}{\overset{\|}{\text{C}}}-\text{C}_6\text{H}_4-\overset{\displaystyle\text{O}}{\overset{\|}{\text{C}}}-\text{O}-(\text{CH}_2)_4-\text{O}-\overset{\displaystyle\text{O}}{\overset{\|}{\text{C}}}-\text{C}_6\text{H}_4-
$$

The polymer is

    A   a polyamide formed by addition polymerisation

    B   a polyamide formed by condensation polymerisation

    C   a polyester formed by addition polymerisation

    D   a polyester formed by condensation polymerisation.

16. A solution containing two carbohydrates was tested as shown.

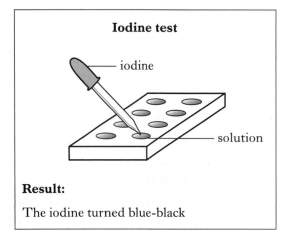

Iodine test

    — iodine

    — solution

Result:

The iodine turned blue-black

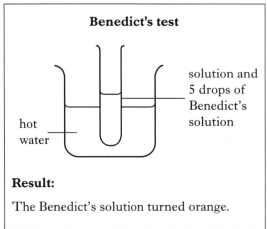

Benedict's test

    solution and 5 drops of Benedict's solution

    hot water

Result:

The Benedict's solution turned orange.

The solution could have contained

    A   starch and sucrose

    B   starch and glucose

    C   glucose and maltose

    D   fructose and sucrose.

17. Which of the following correctly shows the elements present in carbohydrates and proteins?

	Carbohydrates	Proteins
A	C and H	C, H and O
B	C and H	C, H, O and N
C	C, H and O	C, H, O and N
D	C, H, O and N	C, H and O

[Turn over

18. Ammonia dissolves in water to produce a solution with a pH of

A   1

B   4

C   7

D   11.

19. An acidic solution contains

A   only hydrogen ions

B   equal numbers of hydrogen and hydroxide ions

C   more hydrogen ions than hydroxide ions

D   more hydroxide ions than hydrogen ions.

20.

$$ZnO(s) + 2HNO_3(aq) \rightarrow Zn(NO_3)_2(aq) + H_2O(\ell)$$

What type of chemical reaction is represented by the above equation?

A   Condensation

B   Dehydration

C   Neutralisation

D   Precipitation

21. Dilute ethanoic acid exists as an equilibrium mixture.

ethanoic acid molecules $\rightleftharpoons$ ethanoate ions $+$ hydrogen ions

Which statement about the mixture is correct?

A   The ethanoic acid molecules have stopped dissociating into ions.

B   The ethanoic acid molecules have all dissociated into ions.

C   The concentrations of ethanoate ions and hydrogen ions are equal.

D   The concentrations of ethanoic acid molecules and ethanoate ions are equal.

22. Which of the following **does not happen** during the corrosion of iron?

A   The iron is reduced.

B   A compound is formed.

C   Iron(II) ions lose one electron to become iron(III) ions.

D   Iron atoms lose two electrons to form iron(II) ions.

23. In which experiment would the iron nail **not** rust?

A

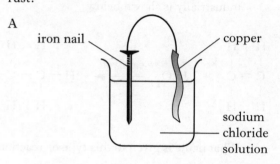

B

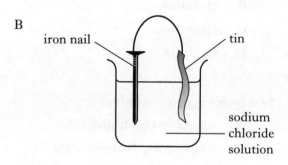

C

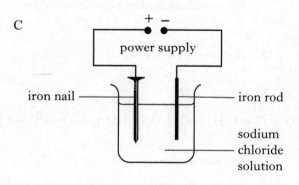

D

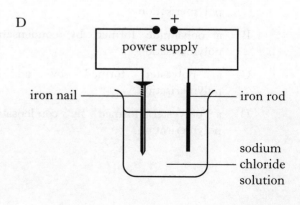

24. Which solution will react with magnesium metal?

You may wish to refer to page 7 of the Data Booklet.

A   Magnesium chloride

B   Potassium chloride

C   Sodium chloride

D   Zinc chloride

25. Which pair of metals, when connected in a cell, would give the highest voltage and a flow of electrons from X to Y?

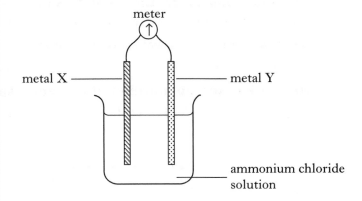

	Metal X	Metal Y
A	magnesium	copper
B	copper	magnesium
C	zinc	tin
D	tin	zinc

[Turn over

SECTION A

PART 2

In Questions 26, 27 and 28 of this part of the paper, an answer is given by circling the appropriate letter (or letters) in the answer grid provided.

In some questions, two letters are required for full marks.

If more than the correct number of answers is given, marks will be deducted.

In some cases, the number of correct responses may NOT be identified in the question.

A total of 5 marks is available in this part of the paper.

SAMPLE QUESTION

A CH_4	B H_2	C CO_2
D CO	E C_2H_5OH	F C

(a) Identify the hydrocarbon.

Ⓐ	B	C
D	E	F

The one correct answer to part (a) is A. This should be circled.

(b) Identify the **two** elements.

A	Ⓑ	C
D	E	Ⓕ

As indicated in this question, there are **two** correct answers to part (b). These are B and F. Both answers are circled.

(c) Identify the substance(s) which can burn to produce **both** carbon dioxide and water.

Ⓐ	B	C
D	Ⓔ	F

There are **two** correct answers to part (c). These are A and E. Both answers are circled.

If, after you have recorded your answer, you decide that you have made an error and wish to make a change, you should cancel the original answer and circle the answer you now consider to be correct. Thus, in part (a), if you want to change an answer A to an answer D, your answer sheet would look like this:

A̶	B	C
Ⓓ	E	F

If you want to change back to an answer which has already been scored out, you should enter a tick (✓) in the box of the answer of your choice, thus:

✓A̶	B	C
D̶	E	F

26. Ions are formed when atoms lose or gain electrons.

A	B
fluorine	oxygen
C	D
potassium	sulphur

Which **two** elements form ions with the same electron arrangement as **argon atoms**?

A	B
C	D

27. Caesium forms a compound with fluorine.

Which statement(s) can be applied to this compound?

A	The formula is CsF_2
B	The bonding is covalent
C	The melt conducts electricity
D	The solid conducts electricity
E	The melting point is higher than $0\,°C$

A
B
C
D
E

[Turn over

Page nine

28. The grid shows some sulphur compounds.

A	B
$CH_3 - S - C_2H_5$	$C_2H_5 - S - C_2H_5$

C	D
$\begin{array}{c} CH_3 \\ \vert \\ CH_3 - C - S - H \\ \vert \\ H \end{array}$	$\begin{array}{c} CH_2 - CH_2 \\ \vert \qquad \vert \\ CH_2 \quad CH_2 \\ \backslash \quad / \\ S \end{array}$

(a) Identify the **two** compounds which have the same molecular formula.

A	B
C	D

(b) Identify the compound which has the general formula $C_nH_{2n}S$.

A	B
C	D

Candidates are reminded that the answer sheet MUST be returned INSIDE this answer book.

Marks

SECTION B

50 marks are available in this section of the paper.

1. The most common isotope of potassium is $_{19}^{39}\mathbf{K}$.

   (a) Complete the table to show the number of particles in an atom of $_{19}^{39}\mathbf{K}$.

Type of particle	Number of particles
proton	
neutron	
electron	

   1

   (b) How do isotopes of potassium differ from each other?

   _____

   _____ 1

   (2)

   [Turn over

Marks

2. A very hot flame is produced when ethyne gas (C_2H_2) burns in a plentiful supply of oxygen.

(a) Name the products formed in this reaction.

_____   1

(b) Ethyne is the first member of an homologous series called the alkynes. Ethyne has the following structure.

$$H-C \equiv C-H$$

All the members of the alkyne series have a triple carbon to carbon bond.

Complete the table below.

Position in series	Name	Molecular formula
1st	ethyne	C_2H_2
2nd		C_3H_4
3rd	butyne	

2

(3)

Page twelve

DO NOT WRITE IN THIS MARGIN

Marks

3. Poly(ethenyl ethanoate) is an addition polymer. Part of its structure is shown below.

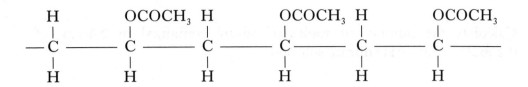

(*a*) Draw the structural formula for the monomer used to make this polymer. **1**

(*b*) Poly(ethenyl ethanoate) is used to make the polymer poly(ethenol).

poly(ethenyl ethanoate) + methanol → poly(ethenol) + methyl ethanoate

Methyl ethanoate is also formed in the reaction.

(i) What property of poly(ethenol) makes it useful for laundry bags?

____

____ **1**

(ii) Draw a structural formula for methyl ethanoate. **1**

(3)

[Turn over

DO NOT WRITE IN THIS MARGIN

Marks

4. (*a*) Write the formula for copper(II) nitrate. **1**

(*b*) Calculate the number of moles of solute contained in $250\,cm^3$ of $0{\cdot}2\ mol\,l^{-1}$ copper(II) nitrate solution.

Answer ____________________ moles **1**

(2)

Marks

5. Students set up the apparatus below to carry out the **PPA "Cracking"**.

mineral wool soaked
in liquid paraffin

powdered aluminium oxide catalyst

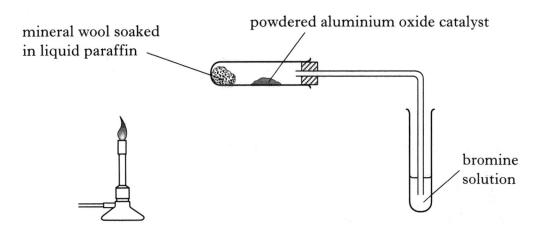

bromine
solution

(a) The PPA gives special instructions as to how the test tube should be heated.

Describe how the test tube should be heated.

_____

_____

_____

_____ **1**

(b) (i) The bromine solution is decolourised.

What does this tell you about the gases produced by cracking?

_____

_____ **1**

(ii) How is "suck back" avoided?

_____

_____ **1**

(c) In this experiment, the aluminium oxide is acting as a heterogeneous catalyst.

What is meant by a **heterogeneous** catalyst?

_____

_____ **1**

(4)

[Turn over

Marks

6. (a) What type of chemical reaction takes place when a metal is obtained from a metal oxide?

    _____

    1

    (b) Name a metal which can be obtained from its oxide using only heat.

    _____

    1

    (c) Iron is produced from iron oxide in a Blast Furnace.

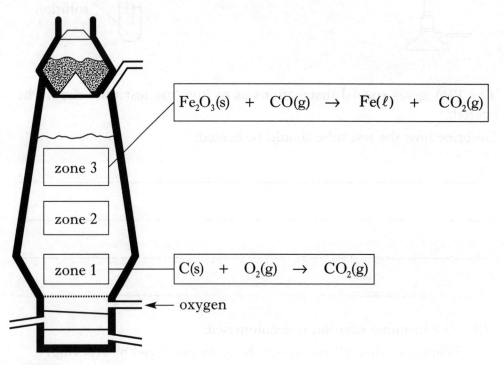

$$Fe_2O_3(s) \ + \ CO(g) \ \rightarrow \ Fe(\ell) \ + \ CO_2(g)$$

$$C(s) \ + \ O_2(g) \ \rightarrow \ CO_2(g)$$

oxygen

Blast Furnace

In zone 1, the coke (carbon) initially reacts with oxygen to produce carbon dioxide.

In zone 3, the iron oxide in the ore reacts with carbon monoxide to produce molten iron metal.

(i) How is the carbon dioxide converted to carbon monoxide in zone 2?

_____

_____

1

(ii) The equation for the reaction taking place in zone 3 is shown below.

$$Fe_2O_3(s) \ + \ CO(g) \rightarrow \ Fe(\ell) \ + \ CO_2(g)$$

Balance this equation.

1

(4)

Marks

7. A copper chloride solution was electrolysed as shown.

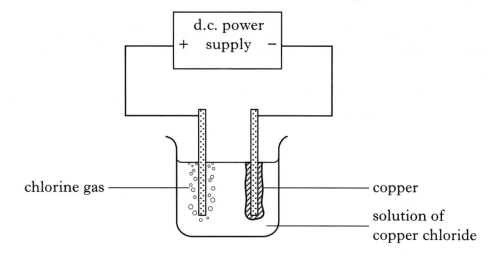

(*a*) Why would a d.c. supply have been used?

_____

_____ 1

(*b*) The chlorine gas can be identified using pH paper or blue litmus paper. Describe how chlorine gas affects pH paper or blue litmus paper.

_____

_____ 1

(2)

[Turn over

DO NOT
WRITE IN
THIS
MARGIN

Marks

8. **A dilute solution has a lower concentration of dissolved substance than a concentrated solution.**

Describe an experiment which could be used to show that 0.1 mol l^{-1} sodium chloride solution has less dissolved salt than 0.2 mol l^{-1} sodium chloride solution.

You may wish to use some or all of the apparatus shown below. You may use other apparatus if required.

_____

_____

_____

_____

_____

_____

_____

_____ (2)

Marks

9. (*a*)   (i)   Describe the test for oxygen gas.

_____

_____   **1**

   (ii)   Explain why this test is not positive with air.

_____

_____   **1**

(*b*)   The astronauts in a space station require a constant supply of oxygen.

In emergencies, oxygen can be produced by decomposing lithium perchlorate ($LiClO_4$).

The equation for the decomposition is shown below.

$$LiClO_4(s) \quad \rightarrow \quad LiCl(g) \quad + \quad 2O_2(g)$$

Calculate the mass of oxygen produced when 1000 g of lithium perchlorate completely decomposes.

Answer = _____ g   **2**

(4)

[Turn over

Marks

10. Enzymes are biological catalysts. A chemistry class was investigating how the activity of an enzyme changed with pH.

They carried out experiments at different pH values and timed how long it took for the enzyme to break down starch. The more active the enzyme, the shorter the time taken to break down the starch.

Here are their results.

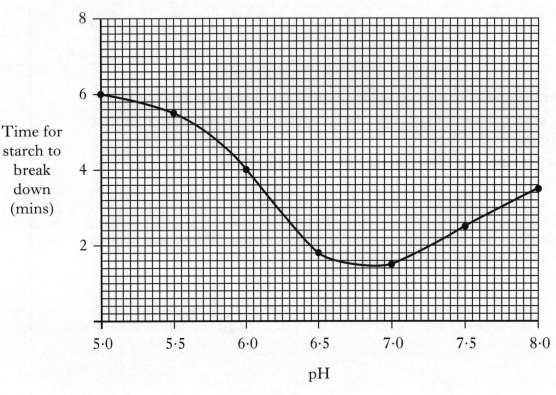

Time for starch to break down (mins)

pH

(a) Name the type of chemical reaction taking place when starch is broken down.

_____   1

(b) To which class of compounds do enzymes belong?

_____   1

(c) What effect does increasing the pH from 5 to 6 have on the **activity of the enzyme**?

_____

_____   1

Marks

10. **(continued)**

(*d*) Some students suggested speeding up the rate of the reaction by heating the starch and enzyme mixtures in boiling water baths.

Why would this **not** speed up the reaction rate?

_____

_____ 1

(*e*) During the experiment, small samples were removed and tested with iodine solution to see if all the starch had been broken down.

Suggest how the students would have known that all the starch had been broken down.

_____

_____ 1

(5)

[Turn over

Marks

11. Hydrochloric acid is a strong acid. The table shows the pH values of some hydrochloric acid solutions.

Concentration (mol l^{-1})	1·0	0·1	0·01	0·001
pH	0	1	2	3

(a) Predict the concentration of hydrochloric acid solution with a pH of 5.

_____ mol l^{-1}

1

(b) Ethanoic acid is a weak acid. The pH of a 0·1 mol l^{-1} solution of ethanoic acid is 3.

Why does this solution have a higher pH than a 0·1 mol l^{-1} solution of hydrochloric acid?

_____

_____

1

(c) Describe an experiment which you could carry out to compare the rate of reaction of hydrochloric acid and ethanoic acid with magnesium.

_____

_____

_____

_____

_____

_____

_____

2

(4)

Marks

12. Metal salts can be made by using different methods.

(*a*) Barium sulphate can be made by reacting solutions of barium chloride and sodium sulphate.

The ionic equation for this reaction is:

$$Ba^{2+}(aq) + 2Cl^-(aq) + 2Na^+(aq) + SO_4^{2-}(aq) \rightarrow Ba^{2+}SO_4^{2-}(s) + 2Na^+(aq) + 2Cl^-(aq)$$

    (i) Rewrite the equation omitting spectator ions.

                                                  1

    (ii) Name the type of reaction taking place.

       _____   1

(*b*) Potassium sulphate can be made by titrating sulphuric acid with potassium hydroxide solution.

$$2KOH(aq) + H_2SO_4(aq) \rightarrow K_2SO_4(aq) + 2H_2O(\ell)$$

If $12 \cdot 5 \, cm^3$ of dilute sulphuric acid were required to neutralise $20 \, cm^3$ of $0 \cdot 1 \, mol \, l^{-1}$ potassium hydroxide solution, calculate the concentration of the sulphuric acid.

                                                  2

(4)

[Turn over

DO NOT
WRITE I
THIS
MARGI

Marks

13. Human blood contains Fe^{3+} ions.

    To find the concentration of Fe^{3+} in blood, the Fe^{3+} ions are first converted into Fe^{2+} ions.

    (a) Write an ion-electron equation for this change.

    _____   1

    (b) Fe^{2+} ions are then reacted to form a purple compound.

    When the solution is placed in a beam of light only some of the light is transmitted.

    These results show the percentage of light transmitted for a series of solutions of known Fe^{2+} concentration.

Concentration of Fe^{2+} (mg l^{-1})	1·0	2·0	3·0	4·0	6·0	8·0
% transmittance	79	65	55	44	27	18

   (i) Plot the results as a line graph.

   (Additional graph paper, if required, can be found on page 28.)

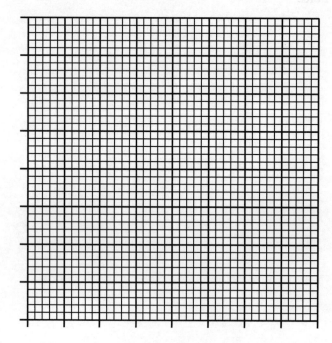

2

   (ii) Using your graph, estimate the concentration of Fe^{2+} ions present in a solution with a transmittance of 36%.

   _____ mg l^{-1} 1

   (4)

Marks

14. A fuel cell which can supply electricity to run a car uses hydrogen and oxygen. The electrodes are made of carbon cloth which is coated with finely divided platinum. The electrodes press against a solid polymer electrolyte.

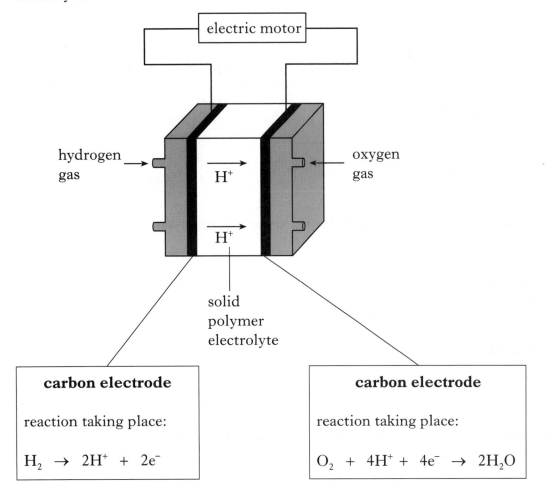

electric motor

hydrogen gas $\rightarrow$ H^+ $\rightarrow$ $\leftarrow$ oxygen gas

H^+ $\rightarrow$

solid polymer electrolyte

carbon electrode

reaction taking place:

$H_2 \rightarrow 2H^+ + 2e^-$

carbon electrode

reaction taking place:

$O_2 + 4H^+ + 4e^- \rightarrow 2H_2O$

(*a*) Combine the 2 ion-electron equations for the electrode reactions to produce a **balanced** REDOX equation.

_____   1

(*b*) Although the polymer is solid, it acts as an electrolyte.

What is an electrolyte?

_____   1

(*c*) The platinum catalyst has been finely divided to increase the surface area.

Why does increasing the surface area increase the rate of reaction?

_____   1

(3)

[Turn over

Marks

15. Hydroxy acids are compounds that contain both a hydroxyl group and a carboxylic acid group within the same molecule.

These compounds are able to form cyclic esters called lactones.

eg

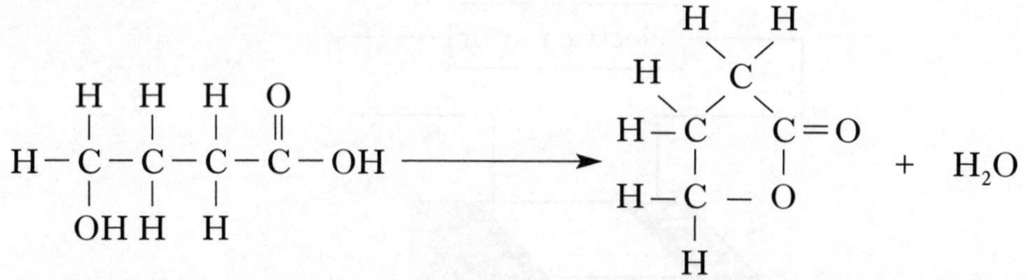

$$H-\underset{\underset{OH}{|}}{\overset{\overset{H}{|}}{C}}-\underset{\underset{H}{|}}{\overset{\overset{H}{|}}{C}}-\underset{\underset{H}{|}}{\overset{\overset{H}{|}}{C}}-\overset{\overset{O}{\|}}{C}-OH \longrightarrow$$

hydroxy acid   a lactone

(a) Draw the structural formula for the lactone formed when this hydroxy acid reacts.

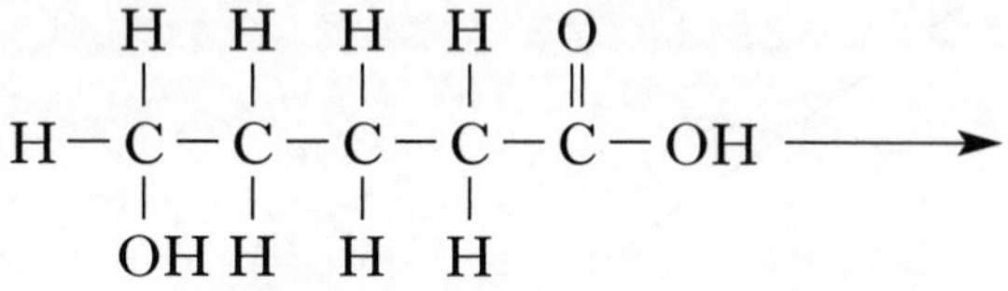

$$H-\underset{\underset{OH}{|}}{\overset{\overset{H}{|}}{C}}-\underset{\underset{H}{|}}{\overset{\overset{H}{|}}{C}}-\underset{\underset{H}{|}}{\overset{\overset{H}{|}}{C}}-\underset{\underset{H}{|}}{\overset{\overset{H}{|}}{C}}-\overset{\overset{O}{\|}}{C}-OH \longrightarrow$$

1

(b) Draw the structural formula for the hydroxy acid from which this lactone was formed.

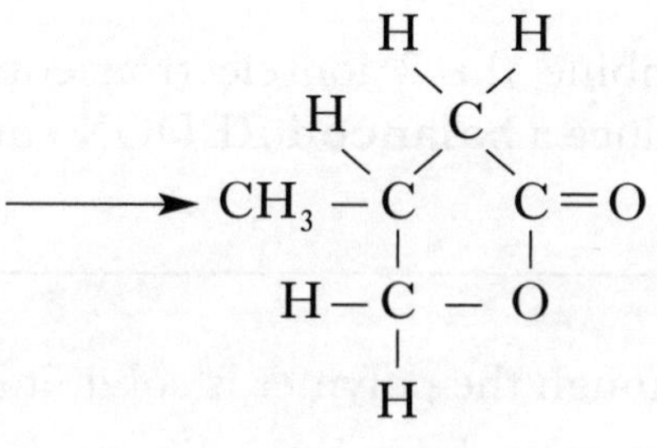

1

(2)

Marks

16.

INTERMEDIATE 2 CHEMISTRY	Reaction of Metals with Oxygen	Unit 3 PPA 3

Aim: The aim of this experiment is to place zinc, copper and magnesium in order of reactivity by observing the ease with which they react with oxygen.

Requirements: Samples of metals, potassium permanganate, mineral wool, dry test tubes, clamp stand and clamp, bunsen.

(*a*) Complete and label the diagram of the test tube to show how it would be set up to burn a metal in oxygen.

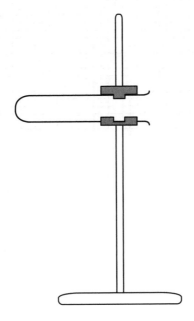

1

(*b*) During the experiment, safety goggles must be worn.

State **one** other safety precaution which must be taken when heating the test tube.

_____

_____   1

(2)

[END OF QUESTION PAPER]

DO NOT
WRITE
THIS
MARGIN

ADDITIONAL SPACE FOR ANSWERS

ADDITIONAL GRAPH PAPER FOR QUESTION 13(b)(i)

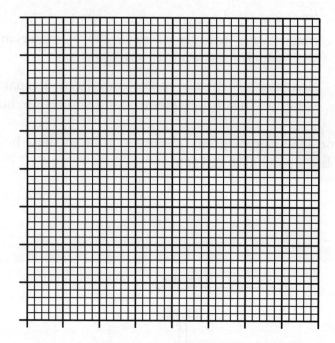

[BLANK PAGE]

FOR OFFICIAL USE

Section B **Total Marks**

X012/201

NATIONAL
QUALIFICATIONS
2002

TUESDAY, 4 JUNE
9.00 AM – 11.00 AM

CHEMISTRY
INTERMEDIATE 2

Fill in these boxes and read what is printed below.

Full name of centre

Town

Forename(s)

Surname

Date of birth
Day Month Year

Scottish candidate number

Number of seat

Necessary data will be found in the Chemistry Data Booklet for Standard Grade and Intermediate 2 (1999 Edition).

Section A —Part 1 Questions 1 to 25 and Part 2 Questions 26 and 27

Instructions for completion of **Part 1** and **Part 2** are given on pages two and seven respectively.

Section B (Questions 1 to 14)

All questions should be attempted.

The questions may be answered in any order but all answers are to be written in the spaces provided in this answer book, and must be written clearly and legibly in ink.

Rough work, if any should be necessary, as well as the fair copy, is to be written in this book.

Rough work should be scored through when the fair copy has been written.

Additional space for answers and rough work will be found at the end of the book. If further space is required, supplementary sheets may be obtained from the invigilator and should be inserted inside the **front** cover of this book.

Before leaving the examination room you must give this book to the invigilator. If you do not, you may lose all the marks for this paper.

SCOTTISH
QUALIFICATIONS
AUTHORITY

SECTION A

PART 1

Read carefully

1. Check that the answer sheet provided is for Chemistry Intermediate 2 (Section A).

2. Fill in the details required on the answer sheet.

3. **In questions 1 to 25 of this part of the paper, an answer is given by indicating the choice A, B, C or D by a stroke made in INK in the appropriate place in Part 1 of the answer sheet—see the sample question below.**

4. **For each question there is only ONE correct answer.**

5. Rough working, if required, should be done only on this question paper, or on the rough working sheet provided—**not** on the answer sheet.

6. At the end of the examination the answer sheet for Section A **must** be placed **inside** the front cover of this answer book.

This part of the paper is worth 25 marks.

SAMPLE QUESTION

To show that the ink in a ball-pen consists of a mixture of dyes, the method of separation would be

　　　　A　fractional distillation

　　　　B　chromatography

　　　　C　fractional crystallisation

　　　　D　filtration.

The correct answer is B—chromatography. A **heavy** vertical line should be drawn joining the two dots in the appropriate box in the column headed **B** as shown **in the example on the answer sheet**.

If, after you have recorded your answer, you decide that you have made an error and wish to make a change, you should cancel the original answer and put a vertical stroke in the box you now consider to be correct. Thus, if you want to change an answer **D** to an answer **B**, your answer sheet would look like this:

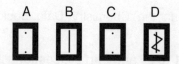

If you want to change back to an answer which has already been scored out, you should **enter a tick (✓)** to the RIGHT of the box of your choice, thus:

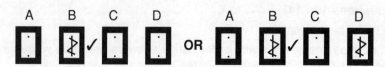

SECTION A

PART 1

1. Which of the following elements is the most recently discovered?

   (You may wish to use page 8 of the data booklet to help you.)

   A   Aluminium

   B   Hydrogen

   C   Iodine

   D   Magnesium

2. Which of the following gases does **not** exist as diatomic molecules?

   A   Nitrogen

   B   Oxygen

   C   Fluorine

   D   Neon

3. An atom has atomic number 17 and mass number 35.

   The number of neutrons in the atom is

   A   17

   B   18

   C   35

   D   52.

4. The formula for magnesium sulphite is

   A MgS

   B $MgSO_3$

   C $MgSO_4$

   D MgS_2O_3.

5. What is the charge on the chromium ion in $CrCl_3$?

   A   1+

   B   1−

   C   3+

   D   3−

6. Which of the following particles contains a different number of electrons from the others?

   (You may wish to use page 1 of the data booklet to help you.)

   A Cl^-

   B S^{2-}

   C Ar

   D Na^+

7. What is the relative formula mass of ammonium sulphate, $(NH_4)_2SO_4$?

   A   70

   B   118

   C   132

   D   228

8. During the electrolysis of copper(II) chloride solution, the reaction taking place at the positive electrode is

   A $Cu^{2+}(aq) + 2e^- \rightarrow Cu(s)$

   B $Cu(s) \rightarrow Cu^{2+}(aq) + 2e^-$

   C $2Cl^-(aq) \rightarrow Cl_2(g) + 2e^-$

   D $Cl_2(g) + 2e^- \rightarrow 2Cl^-(aq)$.

9. The fractional distillation of crude oil depends on the fact that different hydrocarbons have different

   A   densities

   B   solubilities

   C   boiling points

   D   ignition temperatures.

[Turn over

10. Which of the following molecules is an isomer of heptane?

A

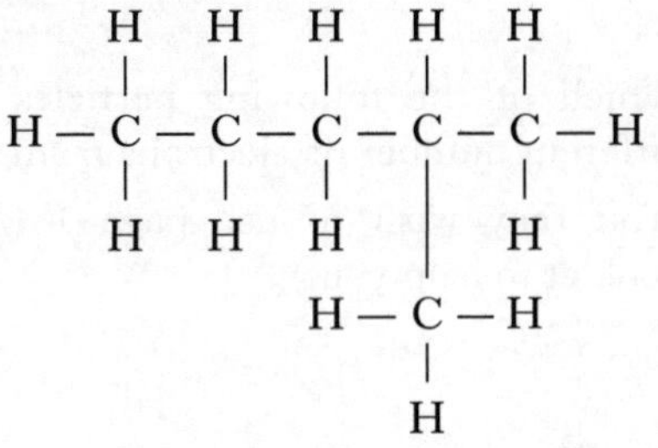

$$H-C-C-C-C-C-H$$

with H atoms and a $H-C-H$ branch (pentane chain with one CH_3 branch)

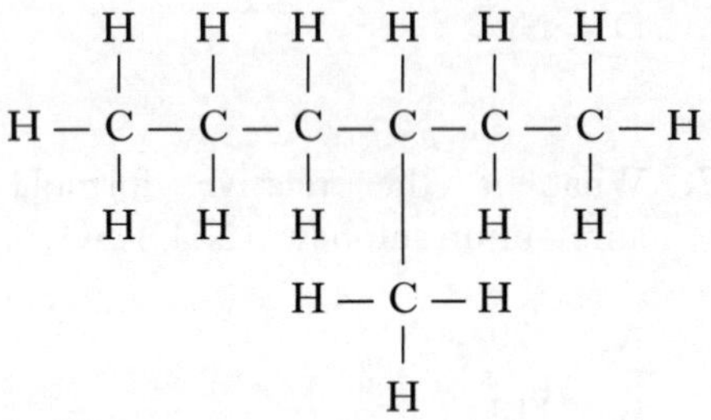

B

$$H-C-C-C-C-C-C-H$$

with H atoms and a $H-C-H$ branch (hexane chain with one CH_3 branch)

C

$$H-C-C-C-C=C-H$$

with a $H-C-H$ branch

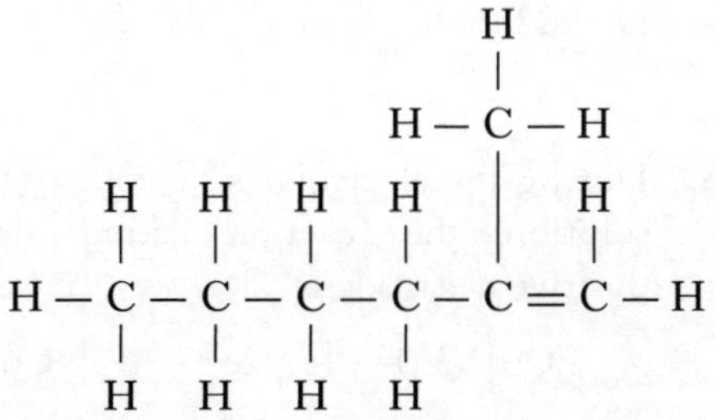

D

$$H-C-C-C-C-C=C-C-H$$

with H atoms

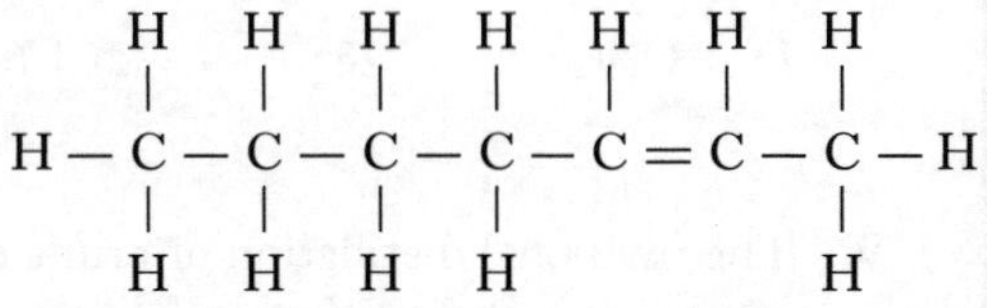

11. Which of the following alcohols has the highest boiling point?

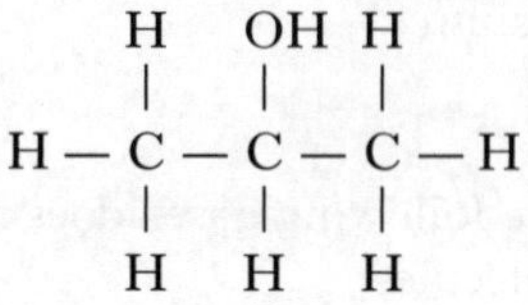

(You may wish to use page 6 of the data booklet to help you.)

A

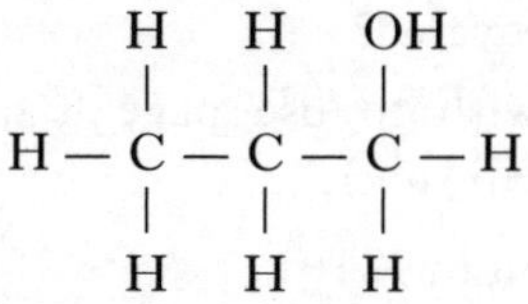

$$H-C-C-C-H$$

with OH on third carbon

B

$$H-C-C-C-H$$

with OH on second carbon

C

$$H-C-C-C-C-H$$

with OH on fourth carbon

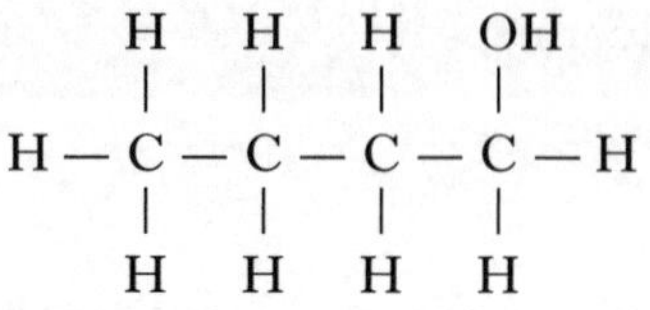

D

$$H-C-C-C-C-H$$

with OH on third carbon

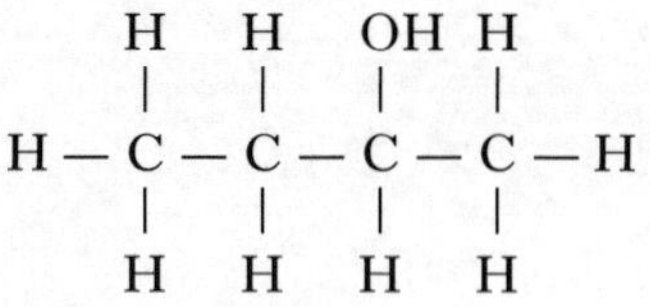

12. Catalytic converters speed up the conversion of harmful gases to less harmful gases. Which of the following reactions is most likely to occur in a catalytic converter?

A   Carbon dioxide reacting to form carbon monoxide

B   Carbon monoxide reacting to form carbon dioxide

C   Nitrogen reacting to form nitrogen dioxide

D   Oxygen reacting to form hydrogen oxide

13. Which of the following are polymers?

A   Plant sugars

B   Animal fats

C   Marine oils

D   Vegetable proteins

14. What type of substance is formed when starch is hydrolysed?

   A   A sugar

   B   A carboxylic acid

   C   An amino acid

   D   An ester

15. What is the ratio of glycerol molecules to fatty acid molecules produced on the hydrolysis of a fat or oil?

   A   1 : 1

   B   1 : 2

   C   1 : 3

   D   1 : 4

16. The apparatus below can be used to dehydrate ethanol.

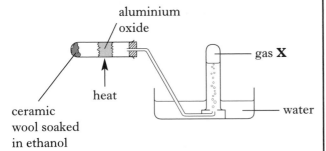

   Gas **X** will

   A   burn with a pop

   B   relight a glowing splint

   C   turn limewater cloudy

   D   rapidly decolourise bromine solution.

17. Which line in the table correctly describes what happens to a dilute solution of hydrochloric acid when water is added to it?

	pH	$H^+(aq)$ concentration
A	increases	increases
B	increases	decreases
C	decreases	increases
D	decreases	decreases

18. Which of the following acid solutions would have the lowest conductivity?

   A $0.1 \, mol \, l^{-1}$ ethanoic acid

   B $0.1 \, mol \, l^{-1}$ hydrochloric acid

   C $0.1 \, mol \, l^{-1}$ nitric acid

   D $0.1 \, mol \, l^{-1}$ sulphuric acid

19. The pH of the solution formed when ammonia is bubbled into water is most likely to be

   A   3

   B   5

   C   7

   D   9.

20. Which of the following statements describes the concentrations of $H^+(aq)$ and $OH^-(aq)$ ions in pure water?

   A   The concentrations of $H^+(aq)$ and $OH^-(aq)$ ions are equal.

   B   The concentrations of $H^+(aq)$ and $OH^-(aq)$ ions are zero.

   C   The concentration of $H^+(aq)$ ions is greater than the concentration of $OH^-(aq)$ ions.

   D   The concentration of $OH^-(aq)$ ions is greater than the concentration of $H^+(aq)$ ions.

21. Which of the following compounds is a base?

   A   Potassium carbonate

   B   Potassium chloride

   C   Potassium nitrate

   D   Potassium sulphate

22. Which of the following sodium chloride solutions would contain most dissolved solute?

   A $100 \, cm^3$ of $4 \, mol \, l^{-1}$ solution

   B $200 \, cm^3$ of $3 \, mol \, l^{-1}$ solution

   C $300 \, cm^3$ of $1 \, mol \, l^{-1}$ solution

   D $400 \, cm^3$ of $0.5 \, mol \, l^{-1}$ solution

[Turn over

23. Some metals can be obtained from their metal oxides by heat alone.

    Which of the following oxides would produce a metal when heated?

    A   Calcium oxide

    B   Copper oxide

    C   Zinc oxide

    D   Silver oxide

24. Which of the following substances will **not** produce a gas when added to dilute hydrochloric acid?

    A   Copper

    B   Zinc

    C   Copper(II) carbonate

    D   Zinc carbonate

25. Which of the following solutions when added to copper(II) chloride solution will produce a precipitate?

    (You may wish to use page 5 of the data booklet to help you.)

    A   Sodium hydroxide solution

    B   Calcium bromide solution

    C   Magnesium nitrate solution

    D   Lithium sulphate solution

SECTION A

PART 2

In Questions 26 and 27 of this part of the paper, an answer is given by circling the appropriate letter (or letters) in the answer grid provided.

In some questions, two letters are required for full marks.

If more than the correct number of answers is given, marks will be deducted.

In some cases, the number of correct responses may NOT be identified in the question.

A total of 5 marks is available in this part of the paper.

SAMPLE QUESTION

A CH_4	B H_2	C CO_2
D CO	E C_2H_5OH	F C

(a) Identify the hydrocarbon.

Ⓐ	B	C
D	E	F

The one correct answer to part (a) is A. This should be circled.

(b) Identify the **two** elements.

A	Ⓑ	C
D	E	Ⓕ

As indicated in this question, there are **two** correct answers to part (b). These are B and F. Both answers are circled.

(c) Identify the substance(s) which can burn to produce **both** carbon dioxide and water.

Ⓐ	B	C
D	Ⓔ	F

There are **two** correct answers to part (c). These are A and E. Both answers are circled.

If, after you have recorded your answer, you decide that you have made an error and wish to make a change, you should cancel the original answer and circle the answer you now consider to be correct. Thus, in part (a), if you want to change an answer A to an answer D, your answer sheet would look like this:

A̶	B	C
Ⓓ	E	F

If you want to change back to an answer which has already been scored out, you should enter a tick (✓) in the box of the answer of your choice, thus:

✓Ⓐ̶	B	C
D̶	E	F

26. The grid shows six oxides.

A	B	C
H_2O	NO_2	K_2O

D	E	F
CaO	CO	SO_2

(a) Identify the oxide produced by the sparking of air in car engines.

(b) Identify the **two** oxides which dissolve in water to produce alkaline solutions.

(c) Identify **two** oxides produced by burning hydrocarbons.

27. Naturally occurring silver (atomic number 47, relative atomic mass 108) consists of a mixture of two isotopes with mass numbers 107 (^{107}Ag) and 109 (^{109}Ag).

Identify the true statement(s).

A	Isotopes of silver have the same number of neutrons.
B	Isotopes of silver have the same number of protons.
C	All silver atoms have a relative atomic mass of 108.
D	Atoms of ^{107}Ag are more abundant than those of ^{109}Ag.
E	All silver atoms have 47 electrons.

Candidates are reminded that the answer sheet for Section A MUST be placed INSIDE the front cover of this answer book.

[Turn over for SECTION B on *Page ten*]

Marks

SECTION B

50 marks are available in this section of the paper.

1. (*a*) Write the formula for barium chloride.

_____

1

(*b*)

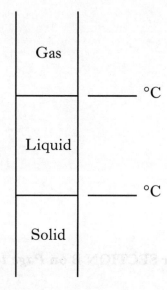

(i) Using information from the data booklet, mark on the diagram the melting point and boiling point of barium chloride.

1

(ii) In what state is barium chloride at 900 °C?

_____

1

(3)

Marks

2. Carborundum and silica are examples of covalent network compounds.

(*a*) In carborundum, the lattice contains **equal** numbers of silicon and carbon atoms.

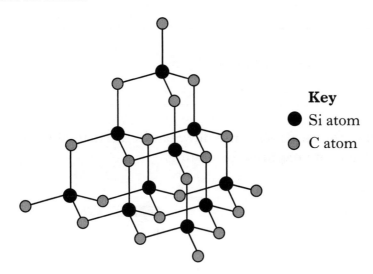

Key
● Si atom
● C atom

Write the formula for carborundum.

_____

1

(*b*) Part of the silica lattice is shown below.

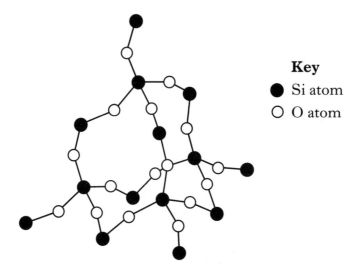

Key
● Si atom
○ O atom

(i) Why does silica **not** conduct electricity?

_____

_____

1

(ii) Why does silica have a high melting point?

_____

_____

1

(3)

DO NO
WRITE
THIS
MARG

Marks

3. (*a*) When iron rusts, iron(II) ions are formed.

Write the ion-electron equation for the formation of iron(II) ions from iron atoms.

_____ **1**

(*b*) Name the solution used to test for iron(II) ions.

_____ **1**

(*c*) The apparatus and chemicals shown below can be used to show that both oxygen and water are required for rusting.

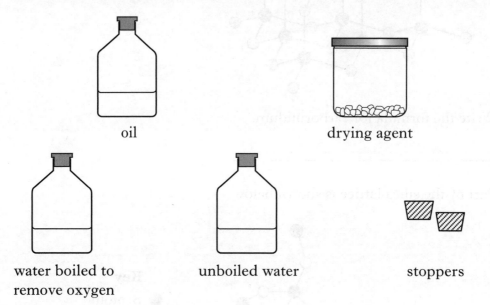

oil   drying agent

water boiled to   unboiled water   stoppers
remove oxygen

The following test tubes were set up.

Experimental conditions		
water and oxygen present	only oxygen present	only water present
test tube **1**	test tube **2**	test tube **3**

Complete and label the diagram of test tube **3**.   **1**

(3)

Marks

4. A new air bag is being developed for use in cars.

In the reaction, butane reacts with an oxide of nitrogen.

$$C_4H_{10} \quad + \quad N_2O \quad \rightarrow \quad CO_2 \quad + \quad H_2O \quad + \quad N_2$$

(*a*) Balance this equation. **1**

(*b*) Nitrogen is formed in the reaction. Draw a diagram to show how the outer electrons are shared in a molecule of nitrogen.

1
(2)

[Turn over

Marks

5. Lavender flowers contain an oil.

(a) Lavender oil is produced from the flowers by steam distillation. The flowers are put into a flask with a little water. Steam from a steam generator is blown through them to extract the oil. The mixture of lavender oil and steam distils over. It is condensed and collected.

The pieces of apparatus which are used to carry out this steam distillation are shown below.

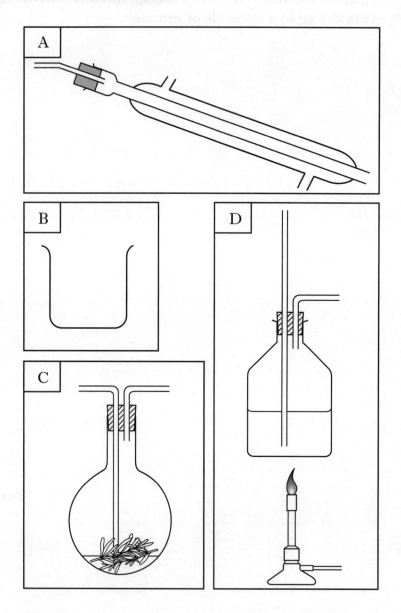

Put a letter in each box to show the order in which the pieces of apparatus should be arranged to obtain the mixture.

1

Marks

5. (continued)

(b) The structural formulae for two of the compounds in lavender oil are shown below.

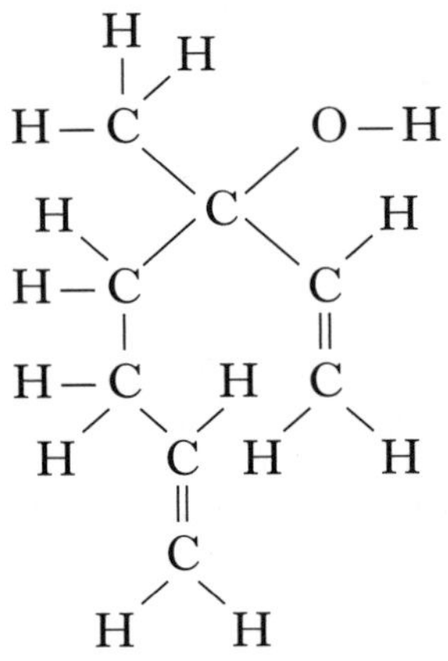

linalool 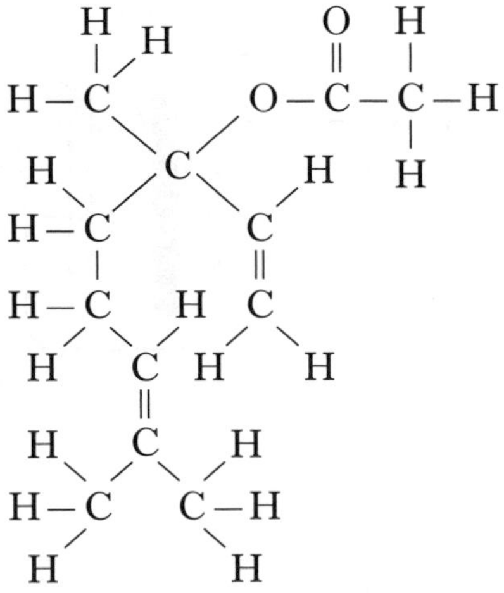   **linalyl ethanoate**

(i) To show that lavender oil contains unsaturated compounds it can be tested by shaking a sample with bromine solution.

Bromine solution is corrosive. Apart from wearing safety goggles, give another safety precaution which should be taken when shaking the sample with bromine solution.

_____

_____   **1**

(ii) Linalool is an alcohol. Circle the alcohol group in the linalool molecule shown above.   **1**

(iii) Linalyl ethanoate is made from an alcohol and ethanoic acid.
To which group of compounds does linalyl ethanoate belong?

_____   **1**

(iv) Draw the full structural formula for ethanoic acid.

1

(5)

Marks

6. In a **PPA**, students were asked to investigate how different metals affect the size of the voltage generated by a simple cell.

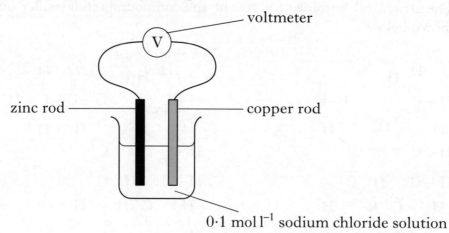

voltmeter

zinc rod

copper rod

$0.1 \, mol \, l^{-1}$ sodium chloride solution

Results:

Metals used	Average voltage (V)
iron and copper	0·5
zinc and copper	
zinc and iron	0·2

(a) What should be done to the metal rods **before** connecting them in the cell?

_____

_____   **1**

(b) State **two** factors which the students would have kept the same during this experiment.

_____

_____

_____   **1**

(c) Complete the results table above by predicting the average voltage reading which could have been obtained using zinc and copper rods.   **1**

(3)

Marks

7. Millions of tonnes of fossil fuels can be saved by burning household rubbish in furnaces to produce energy.

The emissions from these furnaces are carefully controlled to prevent harmful substances being released into the atmosphere.

(a) Nitrogen dioxide gas is produced in the furnaces.

What would form if this gas dissolved in water in the atmosphere?

_____

_____ **1**

(b) Acidic hydrogen chloride is removed from the emissions using calcium hydroxide.

$$2HCl \ + \ Ca(OH)_2 \ \longrightarrow \ CaCl_2 \ + \ 2H_2O$$

Name this type of chemical reaction.

_____ **1**

(c) To remove small ash and soot particles, the emissions are passed through cloth filters coated with the polymer, poly(tetrafluoroethene).

A section of the polymer is shown.

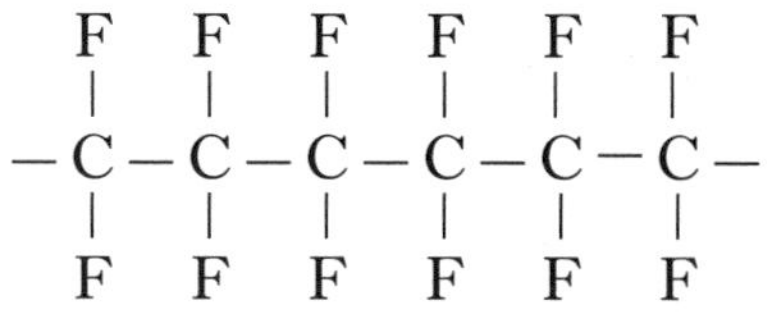

```
    F   F   F   F   F   F
    |   |   |   |   |   |
  — C — C — C — C — C — C —
    |   |   |   |   |   |
    F   F   F   F   F   F
```

Draw the full structural formula for the monomer from which poly(tetrafluoroethene) is made.

1

(3)

[Turn over

Marks

8. Sodium carbonate is an important industrial chemical which is made from sodium chloride by the Solvay Process.

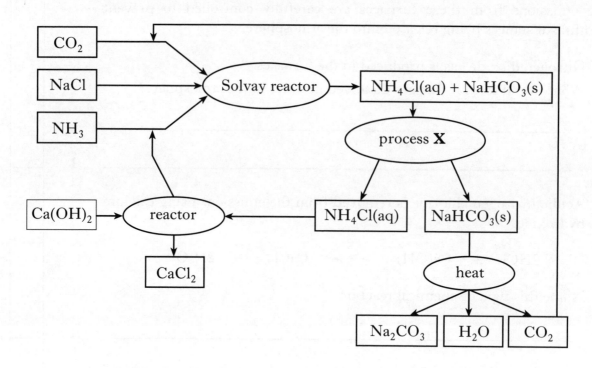

(a) Name process **X**.

_____ 1

(b) Identify a substance which is recycled.

_____ 1

(c) The salt sodium carbonate Na_2CO_3 is the main product of the Solvay process.

Name another **salt** produced.

_____ 1

Marks

8. **(continued)**

(*d*) The main use for sodium carbonate is glassmaking, for which a high purity is required. The purity of a sample of sodium carbonate can be checked by titration with acid.

$$Na_2CO_3(aq) + 2HCl(aq) \longrightarrow 2NaCl(aq) + CO_2(g) + H_2O(\ell)$$

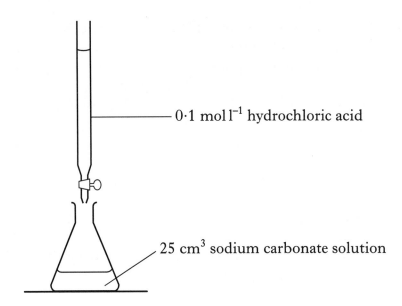

— $0.1 \, mol \, l^{-1}$ hydrochloric acid

— $25 \, cm^3$ sodium carbonate solution

$22.4 \, cm^3$ acid was required in the titration of this sodium carbonate solution.

Calculate the concentration, in $mol \, l^{-1}$, of the sodium carbonate solution.

_____ $mol \, l^{-1}$

2

(5)

[Turn over

Marks

9. Sodium thiosulphate ($Na_2S_2O_3$) reacts with hydrochloric acid as shown.

$$(Na^+)_2S_2O_3{}^{2-}(aq) + 2H^+Cl^-(aq) \longrightarrow 2Na^+Cl^-(aq) + S(s) + SO_2(g) + H_2O(\ell)$$

(a) By omitting spectator ions, write the ionic equation for the reaction.

_____   1

(b) This reaction is used in a **PPA** to study the effect of temperature on reaction rate. The rate is determined from the time taken to produce a certain amount of sulphur.

(i) How would you decide when to stop timing?

_____

_____

_____   1

(ii) The experiment must be carried out in a well ventilated area. Give a reason for this.

_____

_____   1

Marks

9. (continued)

(c) A student investigated the effect of changing the **concentration** of the sodium thiosulphate solution on the reaction rate.

The results obtained are shown.

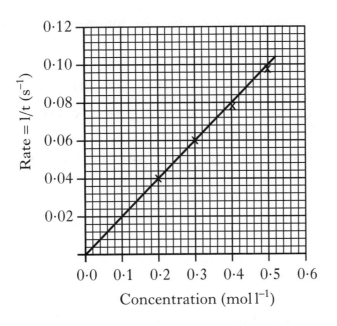

(i) Use the graph to find the time taken, in seconds, when the experiment was carried out using $0.1 \, mol \, l^{-1}$ sodium thiosulphate solution.

_____ s

1

(ii) The graph shows that the reaction rate increases as concentration increases. Use the collision theory to explain why the reaction rate increases.

_____

_____

_____

_____

1

(5)

[Turn over

Marks

10. Kevlar is a recently developed polymer.

(a) State a useful property of Kevlar.

_____

1

(b) The monomers used to make Kevlar have the following structural formulae.

$$HO-\overset{\overset{\displaystyle O}{\|}}{C}-C_6H_4-\overset{\overset{\displaystyle O}{\|}}{C}-OH \qquad H-\overset{\overset{\displaystyle H}{|}}{N}-C_6H_4-\overset{\overset{\displaystyle H}{|}}{N}-H$$

Why are these molecules able to take part in condensation polymerisation?

_____

_____

1

(c) Kevlar is an example of a polyamide.
Draw the structure of an amide link.

1

(3)

Marks

11. (*a*) What is the name given to the group 7 elements?

_____ **1**

(*b*) The diagram shows the chemical energies of the reactants and products when fluorine reacts with hydrogen.

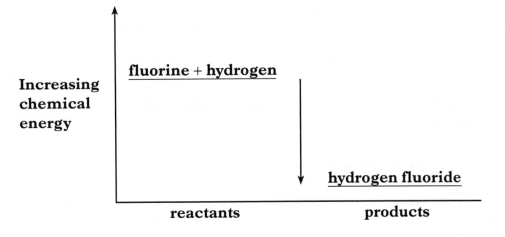

(i) What does the diagram indicate about the chemical reaction?

_____

_____ **1**

(ii) Hydrogen fluoride forms a very weak acid when it is dissolved in water. What is meant by a **weak** acid?

_____

_____ **1**

(3)

[Turn over

Marks

12. Iodine can react with propene in the following way.

$$I_2 \ + \ \underset{\substack{|\\H}}{\overset{\substack{H\\|}}{H-C}} - \underset{\substack{|\\H}}{\overset{\substack{H\\|}}{C}} = C \overset{H}{\underset{H}{\diagup}} \longrightarrow \underset{\substack{|\\H}}{\overset{\substack{H\\|}}{H-C}} - \underset{\substack{|\\H}}{\overset{\substack{I\\|}}{C}} - \underset{\substack{|\\H}}{\overset{\substack{I\\|}}{C}} - H$$

(a) (i) Name the homologous series to which propene belongs.

_____   1

(ii) Name the type of chemical reaction which takes place when iodine reacts with propene.

_____   1

(b) Calculate the mass of iodine, in grams, that will react with 100 g of propene.

Space for working

_____ g   2

(c) The mass of iodine that reacts with 100 g of a substance is known as the iodine number.

Explain why oils are likely to have higher iodine numbers than fats.

_____

_____   1

(d) Liquid oils can be converted into hardened fats using a solid catalyst. The catalyst used is the transition metal, nickel. What type of catalyst is the nickel?

_____   1

(6)

DO NOT
WRITE IN
THIS
MARGIN

Marks

13. (*a*) What mass of copper(II) sulphate (gram formula mass = 159·5 g) is required to make $400 \, cm^3$ of $0·50 \, mol \, l^{-1}$ copper(II) sulphate solution?

2

(*b*) When zinc is added to $0·50 \, mol \, l^{-1}$ copper(II) sulphate solution, the blue colour fades and brown copper metal forms.

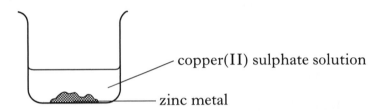

copper(II) sulphate solution

zinc metal

(i) The ion-electron equations for the oxidation and reduction reactions are

$$Zn(s) \longrightarrow Zn^{2+}(aq) + 2e^-$$
$$Cu^{2+}(aq) + 2e^- \longrightarrow Cu(s)$$

Combine the two ion-electron equations to give the redox equation for the reaction.

1

(ii) During the first 60 seconds of the reaction, the concentration of the copper ions drops from $0·50 \, mol \, l^{-1}$ to $0·41 \, mol \, l^{-1}$.

What is the average rate of the reaction, in $mol \, l^{-1} \, s^{-1}$, during the first 60 seconds?

_____ $mol \, l^{-1} \, s^{-1}$ 1

(4)

14. Gases can be liquefied by increasing the pressure, but above a certain temperature it is not possible to do this. This temperature is known as the critical temperature. The critical temperatures of some alkanes are shown below.

Alkane	**Critical temperature (°C)**
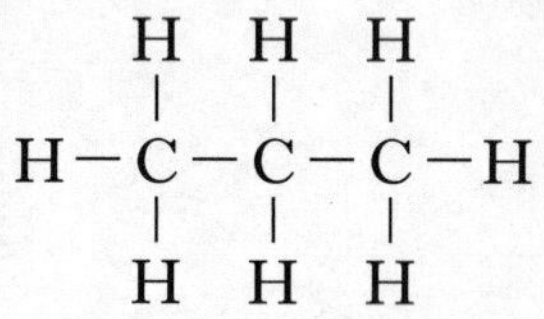	97
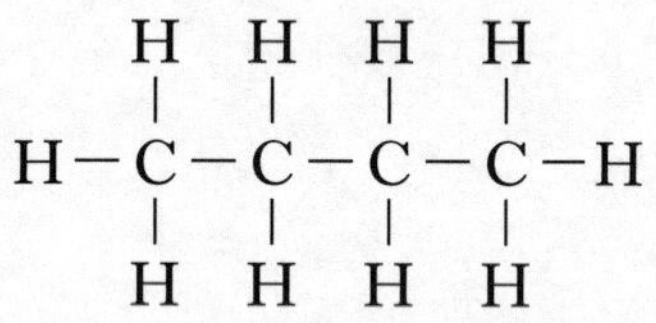	152
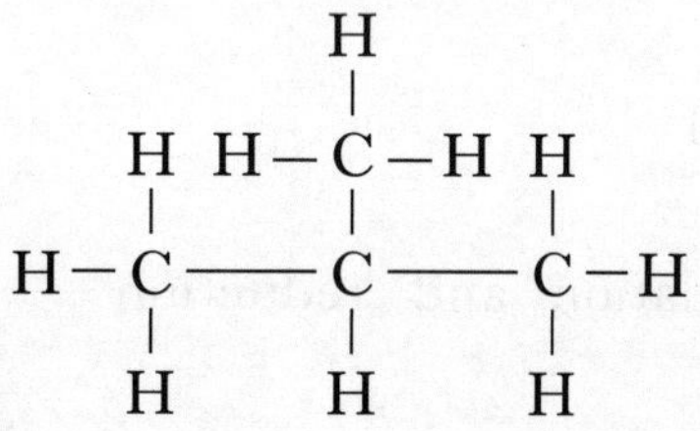	135
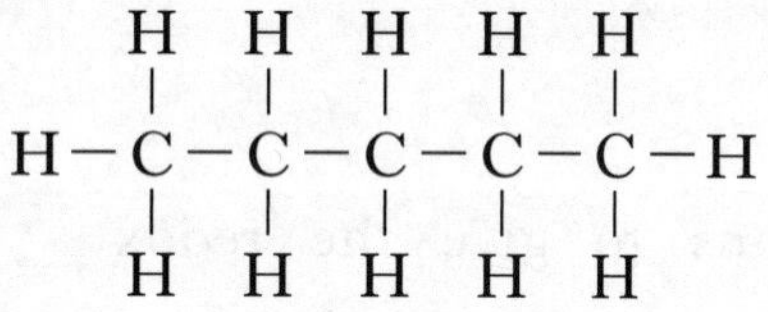	197
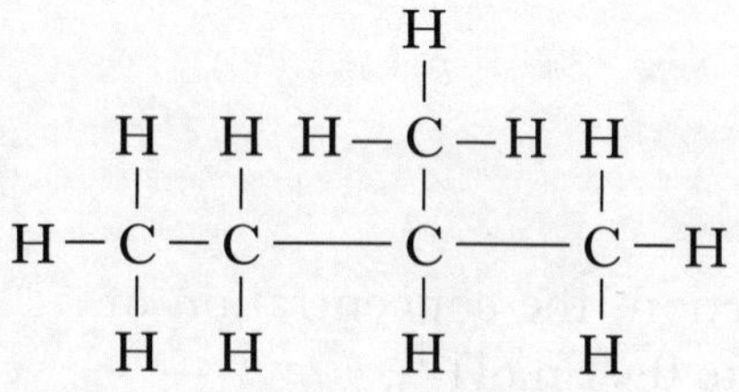	187
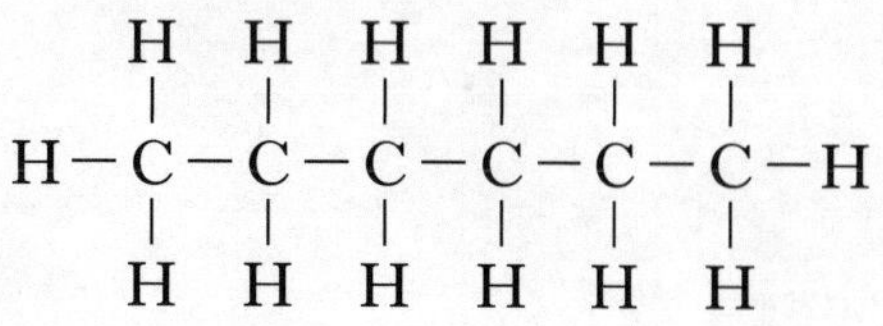	234

Marks

14. (continued)

(a) Describe the trend in critical temperatures for the straight-chain alkanes.

_____

_____   1

(b) Predict the critical temperature of the alkane

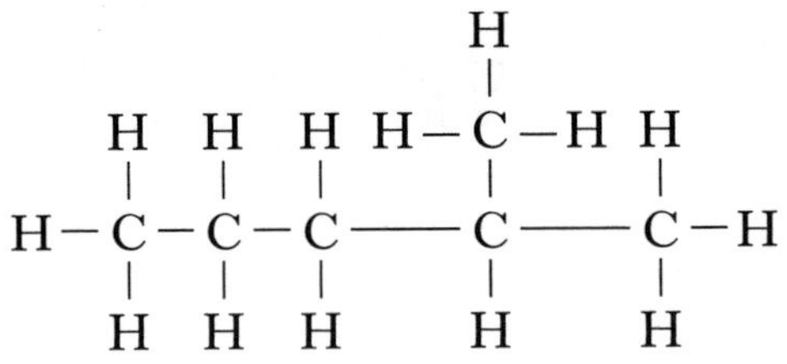

```
                    H
                    |
   H   H   H  H – C – H  H
   |   |   |         |    |
H– C – C – C ——— C ——— C – H
   |   |   |         |    |
   H   H   H         H    H
```
_____ °C   1

                                                (2)

[END OF QUESTION PAPER]

ADDITIONAL SPACE FOR ANSWERS

[BLANK PAGE]

FOR OFFICIAL USE

Section B   **Total Marks**

X012/201

NATIONAL
QUALIFICATIONS
2003

FRIDAY, 23 MAY
1.00 PM – 3.00 PM

CHEMISTRY
INTERMEDIATE 2

Fill in these boxes and read what is printed below.

Full name of centre

Town

Forename(s)

Surname

Date of birth
Day Month Year   Scottish candidate number   Number of seat

Necessary data will be found in the Chemistry Data Booklet for Standard Grade and Intermediate 2 (1999 Edition).

Section A—Questions 1 to 30

Instructions for completion of **Section A** are given on page two.

Section B—Questions 1 to 15

All questions should be attempted.

The questions may be answered in any order but all answers are to be written in the spaces provided in this answer book, and must be written clearly and legibly in ink.

Rough work, if any should be necessary, as well as the fair copy, is to be written in this book.

Rough work should be scored through when the fair copy has been written.

Additional space for answers and rough work will be found at the end of the book. If further space is required, supplementary sheets may be obtained from the invigilator and should be inserted inside the **front** cover of this book.

Before leaving the examination room you must give this book to the invigilator. If you do not, you may lose all the marks for this paper.

SCOTTISH
QUALIFICATIONS
AUTHORITY

SECTION A

Read carefully

1. Check that the answer sheet provided is for Chemistry Intermediate 2 (Section A).

2. Fill in the details required on the answer sheet.

3. **In questions 1 to 30 of this part of the paper, an answer is given by indicating the choice A, B, C or D by a stroke made in INK in the appropriate place of the answer sheet—see the sample question below.**

4. **For each question there is only ONE correct answer.**

5. Rough working, if required, should be done only on this question paper, or on the rough working sheet provided—**not** on the answer sheet.

6. At the end of the examination the answer sheet for Section A **must** be placed **inside** the front cover of this answer book.

This part of the paper is worth 30 marks.

SAMPLE QUESTION

To show that the ink in a ball-pen consists of a mixture of dyes, the method of separation would be

        A fractional distillation

        B chromatography

        C fractional crystallisation

        D filtration.

The correct answer is B—chromatography. A **heavy** vertical line should be drawn joining the two dots in the appropriate box in the column headed **B** as shown **in the example on the answer sheet**.

If, after you have recorded your answer, you decide that you have made an error and wish to make a change, you should cancel the original answer and put a vertical stroke in the box you now consider to be correct. Thus, if you want to change an answer **D** to an answer **B**, your answer sheet would look like this:

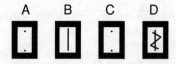

If you want to change back to an answer which has already been scored out, you should **enter a tick (✓)** to the RIGHT of the box of your choice, thus:

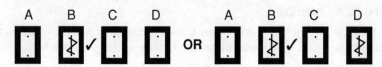

SECTION A

1. Which of the following compounds contains both a transition metal ion and a halide ion?

   A   Aluminium bromide

   B   Cobalt chloride

   C   Iron oxide

   D   Sodium fluoride

2. Magnesium was reacted with dilute hydrochloric acid under different conditions. In each experiment an **excess** of magnesium was added.

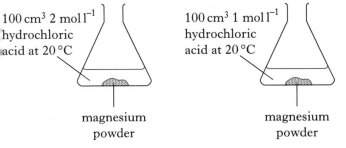

	Rate of reaction	**Volume of gas produced**
A	faster	less
B	faster	the same
C	slower	less
D	slower	the same

   Which line in the table correctly describes **Reaction 2** when compared to **Reaction 1**?

3. An atom is neutral because

   A   the number of protons equals the number of neutrons

   B   the number of electrons equals the number of protons

   C   the number of electrons equals the number of protons plus neutrons

   D   the number of neutrons equals the number of protons plus electrons.

4. Which of the following is the electron arrangement for an atom of an alkali metal?

   A   2,8,1

   B   2,8,2

   C   2,8,3

   D   2,8,4

5. Different isotopes of the same element have identical

   A   nuclei

   B   mass numbers

   C   atomic numbers

   D   numbers of neutrons.

6. A metal **X** reacts with oxygen to form an oxide, X_2O_3.

   During the reaction each atom of metal **X**

   A   gains two electrons

   B   gains three electrons

   C   loses two electrons

   D   loses three electrons.

7. Which of the following elements conducts electricity?

   A   Bromine

   B   Mercury

   C   Oxygen

   D   Sulphur

8. The formula for potassium sulphate is

   A P_2SO_3

   B K_2SO_4

   C P_2SO_4

   D K_2S.

[Turn over

9. Which of the following compounds has an isomer?

A

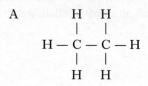

$$H - \overset{\displaystyle H}{\underset{\displaystyle H}{\overset{|}{\underset{|}{C}}}} - \overset{\displaystyle H}{\underset{\displaystyle H}{\overset{|}{\underset{|}{C}}}} - H$$

B

$$\overset{\displaystyle H}{\underset{\displaystyle H}{\overset{|}{\underset{|}{C}}}} = \overset{\displaystyle H}{\underset{\displaystyle H}{\overset{|}{\underset{|}{C}}}}$$

C

$$H - \overset{\displaystyle H}{\underset{\displaystyle H}{\overset{|}{\underset{|}{C}}}} - \overset{\displaystyle H}{\underset{\displaystyle H}{\overset{|}{\underset{|}{C}}}} - \overset{\displaystyle H}{\underset{\displaystyle H}{\overset{|}{\underset{|}{C}}}} - H$$

D

$$\overset{\displaystyle H}{\underset{\displaystyle H}{\overset{|}{\underset{|}{C}}}} = \overset{\displaystyle H}{\underset{\displaystyle H}{\overset{|}{\underset{|}{C}}}} - \overset{\displaystyle H}{\underset{\displaystyle H}{\overset{|}{}}} - H$$

10. Which of the following is a structural formula for methyl ethanoate?

A

$$CH_3 - \overset{\displaystyle O}{\overset{\|}{C}} - O - CH_3$$

B

$$CH_3 - \overset{\displaystyle O}{\overset{\|}{C}} - O - CH_2 - CH_3$$

C

$$CH_3 - CH_2 - \overset{\displaystyle O}{\overset{\|}{C}} - O - CH_3$$

D

$$H - \overset{\displaystyle O}{\overset{\|}{C}} - O - CH_2 - CH_3$$

11. The flow diagram shows the manufacture of polythene from hydrocarbons in crude oil.

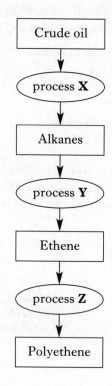

Which line in the table identifies the processes **X**, **Y** and **Z**?

	Process X	Process Y	Process Z
A	distillation	cracking	hydrolysis
B	cracking	combustion	polymerisation
C	polymerisation	distillation	hydrolysis
D	distillation	cracking	polymerisation

12. Ethanol vapour can be dehydrated by passing it over hot aluminium oxide.

Which of the following compounds would be produced?

A   Ethane

B   Ethene

C   Ethanoic acid

D   Ethyl ethanoate

13. Which of the following polymers readily dissolves in water?

A   Poly(ethene)

B   Perspex

C   Poly(ethenol)

D   Kevlar

14. Part of the structure of an addition polymer is shown below. It is made using two different monomers.

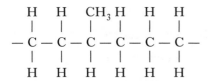

Which pair of alkenes could be used as monomers?

A Ethene and propene

B Ethene and butene

C Propene and butene

D Ethene and pentene

15. Which of the following correctly shows all of the elements always present in carbohydrates and proteins?

	Carbohydrates	Proteins
A	C, H and O	C, H and N
B	C and H	C, H and O
C	C, H, O and N	C and H
D	C, H and O	C, H, O and N

16. Which of the following substances dissolves in water to give a solution of pH greater than 7?

A Ammonia

B Carbon dioxide

C Sulphur dioxide

D Sodium chloride

17. 0·25 mol of potassium hydroxide was dissolved in water and the solution made up to 500 cm^3.

What was the concentration, in mol l^{-1}, of the solution which was formed?

A 0·0005

B 0·125

C 0·5

D 2·0

18. Which of the following statements is true for lithium hydroxide solution?

A It has a pH less than 7.

B It contains no H$^+$(aq) ions.

C It is an example of a strong base.

D It reacts with an acid to form hydrogen gas.

19. Which of the following reacts with dilute hydrochloric acid to give hydrogen gas?

A Copper

B Gold

C Magnesium

D Mercury

Questions 20 and 21 refer to the following reaction.

When lead(II) nitrate solution is added to sodium iodide solution a precipitate of lead(II) iodide is formed.

20. A sample of precipitate can be separated from the mixture by

A condensation

B distillation

C evaporation

D filtration.

21. The equation for the reaction is:

$$Pb^{2+}(aq) + 2NO_3^-(aq) + 2Na^+(aq) + 2I^-(aq)$$
$$\downarrow$$
$$Pb^{2+}(I^-)_2(s) + 2Na^+(aq) + 2NO_3^-(aq)$$

The spectator ions present in this reaction are

A Na$^+$(aq) and NO$_3^-$(aq)

B Na$^+$(aq) and I$^-$(aq)

C Pb^{2+}(aq) and NO$_3^-$(aq)

D Pb^{2+}(aq) and I$^-$(aq).

22. Which of the following metals would react with zinc chloride solution?

(You may wish to use page 7 of the data booklet to help you.)

A Copper

B Gold

C Iron

D Magnesium

23.

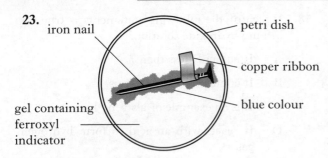

iron nail — petri dish — copper ribbon — blue colour — gel containing ferroxyl indicator

Which ion gives a blue colour with ferroxyl indicator?

A $OH^-(aq)$

B $Fe^{2+}(aq)$

C $Fe^{3+}(aq)$

D $Cu^{2+}(aq)$

24. An atom has 26 protons, 26 electrons and 30 neutrons. The atom will have

A   atomic number 26, mass number 56

B   atomic number 56, mass number 30

C   atomic number 30, mass number 26

D   atomic number 52, mass number 56.

25. Which of the following substances does **not** exist as diatomic molecules?

A   Bromine

B   Carbon monoxide

C   Oxygen

D   Water

26. The properties of fractions obtained from crude oil depend on the sizes of molecules in the fractions.

Compared with a fraction containing small molecules a fraction containing large molecules will

A   be more viscous

B   be more flammable

C   evaporate more readily

D   have a lower boiling point range.

27. During digestion starch is broken down to form glucose.

What name is given to this type of reaction?

A   Combustion

B   Condensation

C   Fermentation

D   Hydrolysis

28. When dilute hydrochloric acid is added to substance **X**, a gas is given off. This gas quickly puts out the candle flame.

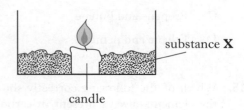

substance **X**

candle

Which of the following could be substance **X**?

A   Calcium hydroxide

B   Calcium carbonate

C   Calcium oxide

D   Calcium

29. The formula mass of ammonium carbonate, $(NH_4)_2CO_3$, is

A   52

B   64

C   96

D   110.

30. Which sugar will **not** be detected by the Benedict's test?

A   Fructose

B   Glucose

C   Maltose

D   Sucrose

Candidates are reminded that the answer sheet for Section A MUST be placed INSIDE the front cover of this answer book.

Marks

SECTION B

50 marks are available in this section of the paper.

1. Yogurt is made by fermenting fresh milk. Enzymes help to convert lactose in the milk to lactic acid.

   (a) What is an enzyme?

   _____   1

   (b) The structural formula for lactic acid is shown below.

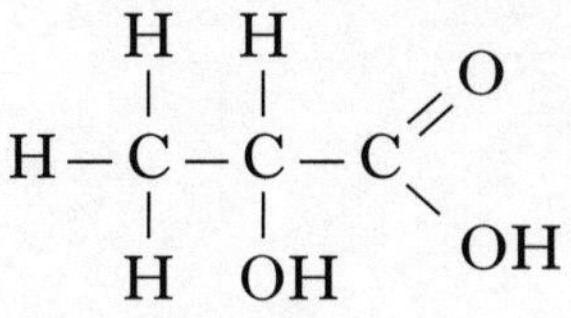

$$H - \overset{\displaystyle H}{\underset{\displaystyle H}{C}} - \overset{\displaystyle H}{\underset{\displaystyle OH}{C}} - C \overset{\displaystyle O}{\underset{\displaystyle OH}{}}$$

   Circle the carboxyl group in the lactic acid molecule.   1

   (c) Sugar can be added after fermentation has taken place to sweeten the yogurt.
   Suggest why the sugar is added after the fermentation stage and not before.

   _____

   _____

   _____   1

   (3)

Marks

2. Dinitrogen monoxide can be used to boost the performance of racing car engines.

When dinitrogen monoxide decomposes it forms a mixture of nitrogen and oxygen. The reaction is exothermic.

$$2N_2O(g) \longrightarrow 2N_2(g) + O_2(g)$$

(*a*) What is meant by an exothermic reaction?

_____

_____ **1**

(*b*) How many moles of oxygen will be produced when four moles of dinitrogen monoxide are decomposed.

_____ **1**

(*c*) An experiment was set up as shown below.

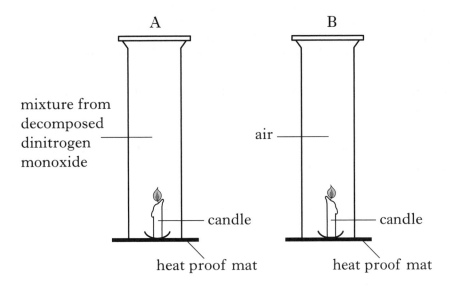

Why will the candle burn for longer in gas jar A?

_____

_____ **1**

(3)

[Turn over

Marks

3. A student set up the following cell.

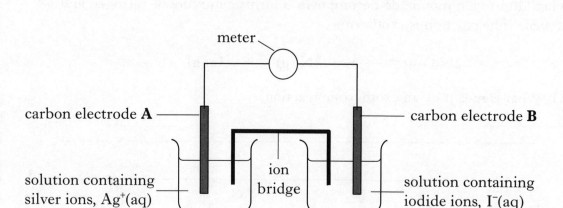

Electrode	Reactions taking place
A	$Ag^+(aq) + e^- \longrightarrow Ag(s)$
B	$2I^-(aq) \longrightarrow I_2(s) + 2e^-$

(a) On the diagram, clearly mark the path and the direction of the electron flow.

1

(b) Combine the two ion-electron equations for the electrode reactions to produce a **balanced** redox equation.

_____

1

(c) What is the purpose of the ion bridge?

_____

1

(d) Describe the chemical test which could be used to show that iodine is formed at electrode **B**.

_____

_____

1

(4)

Marks

4. The table below shows the saturated and unsaturated fatty acid content of a fat and an oil.

Source	Fat/oil	% Fatty acid in substance	
		Saturated	Unsaturated
animal	chicken fat	68	32
marine	cod liver oil	25	75

(a) What do fats and oils provide in our diet?

_____   1

(b) Name another source of fats and oils.

_____   1

(c) Why do oils have a lower melting point than fats?

_____

_____   1

(d) How can oils be converted into hardened fats?

_____

_____   1

(4)

[Turn over

Marks

5. Energy is required to remove an electron from an atom.

The graph shows the energy required to do this for the first 19 elements.

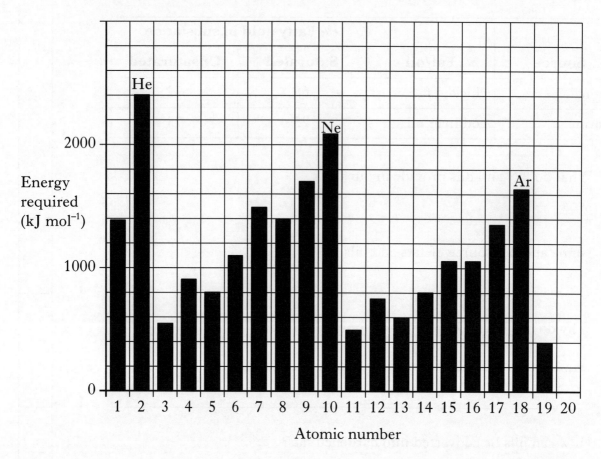

(a) Describe what happens to the energy required going down a group.

_____

_____   1

(b) Describe the general trend in the energy required going from sodium to argon.

_____

_____   1

(c) Draw a bar on the graph to show the energy you would expect to be required for the element with atomic number 20.   1

(3)

Marks

6. Nylon is a polymer.

The monomers shown below are used to produce a nylon.

(*a*) Draw a section of the polymer showing the three monomer units linked together.

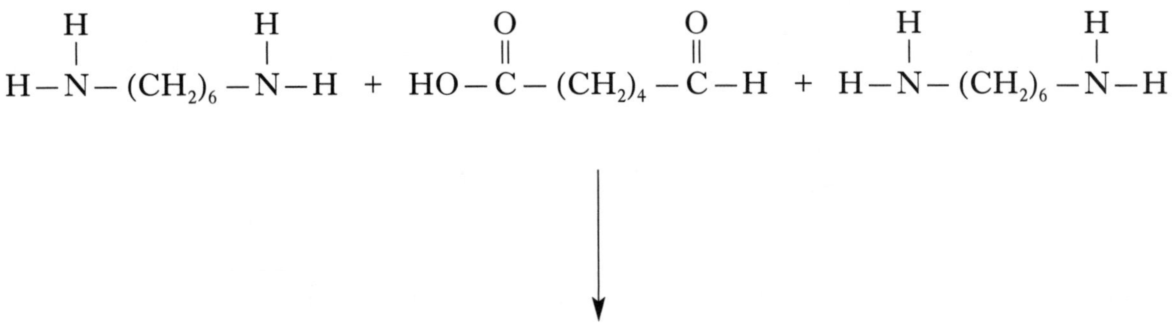

$$\underset{\substack{|\\H}}{H-N}-(CH_2)_6-\underset{\substack{|\\H}}{N}-H \;\; + \;\; HO-\underset{\substack{\|\\O}}{C}-(CH_2)_4-\underset{\substack{\|\\O}}{C}-H \;\; + \;\; \underset{\substack{|\\H}}{H-N}-(CH_2)_6-\underset{\substack{|\\H}}{N}-H$$

1

(*b*) What feature of their structure makes these molecules suitable for use as monomers?

_____

_____ 1

(2)

[Turn over

Marks

7. DIMP is a useful insect repellent.

DIMP contains two ester groups, as shown by the structure below.

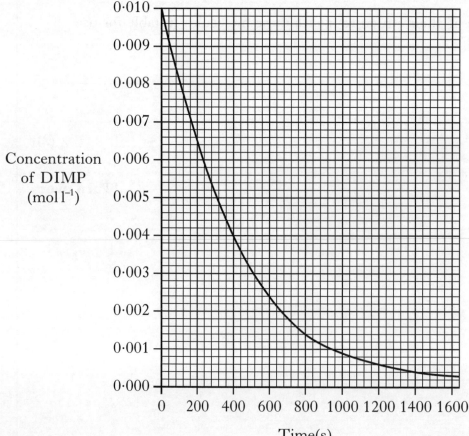

DIMP

When DIMP is hydrolysed it forms a carboxylic acid and an alkanol.

(*a*) Name the alkanol produced when DIMP is hydrolysed.

_____   1

(*b*) An experiment was set up to find out how quickly DIMP is hydrolysed.

The graph shows how the concentration of DIMP changed during hydrolysis.

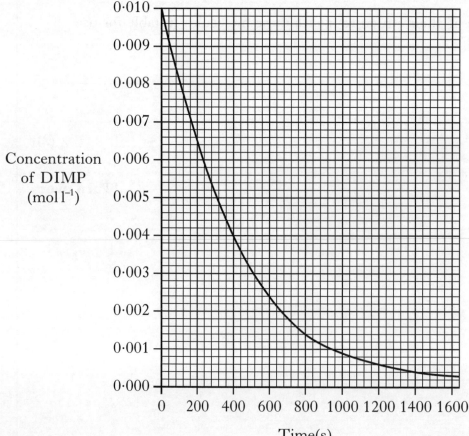

Concentration
of DIMP
$(mol\,l^{-1})$

Time(s)

Marks

7. **(b)** **(continued)**

(i) By how much does the concentration of DIMP fall in the first 400 s?

_____ **1**

(ii) Calculate the average rate of reaction, in $mol\,l^{-1}\,s^{-1}$, between 0 and 400 s.

Space for working

_____ $mol\,l^{-1}\,s^{-1}$ **1**

(3)

[Turn over

Marks

8. (a) In a solution of ethanoic acid not all the molecules break up to form ions.

   What does this indicate about ethanoic acid?

   _____   **1**

   (b) In a solution of hydrochloric acid all of the hydrogen chloride molecules have broken up to form ions.

   Two properties of solutions of ethanoic acid and hydrochloric acid were compared. The actual results for ethanoic acid are shown in the table.

   Circle the appropriate words in the right-hand column of the table to show how the results for hydrochloric acid would compare with those for ethanoic acid.

	$0{\cdot}1$ mol$\,$l^{-1} ethanoic acid	$0{\cdot}1$ mol$\,$l^{-1} hydrochloric acid	
pH	4	lower	higher
Rate of reaction with magnesium	slow	slower	faster

   2

   (c) Vinegar is dilute ethanoic acid. The concentration of ethanoic acid in vinegar can be determined by neutralising a sample of vinegar with sodium hydroxide solution.

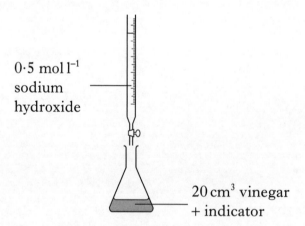

   $0{\cdot}5$ mol$\,$l^{-1} sodium hydroxide

   20 cm^3 vinegar + indicator

   Volume of $0{\cdot}5$ mol$\,$l^{-1} sodium hydroxide solution required $= 33{\cdot}4$ cm^3

DO NOT WRITE IN THIS MARGIN

Marks

8. (*c*) (continued)

The equation for the reaction is

$CH_3COOH(aq) + NaOH(aq) \rightarrow CH_3COONa(aq) +$

Calculate the concentration, in $mol\,l^{-1}$, of the ethanoic acid in vinegar.

______________ $mol\,l^{-1}$ **2**

(5)

[Turn over

Marks

9. Part of a student's PPA instruction sheet is shown below.

Intermediate 2 Chemistry	**Preparation of a Salt**	Unit 3 PPA 1

Aim

The aim of this experiment is to make a magnesium salt by the reaction of magnesium/magnesium carbonate with sulphuric acid.

Procedure

1. Using a measuring cylinder add $20\,cm^3$ of dilute acid to the beaker.

2. Add a spatulaful of magnesium or magnesium carbonate to the acid and stir the reaction mixture with a glass rod.

3. If all the solid reacts add another spatulaful of magnesium or magnesium carbonate and stir the mixture.

4. Continue adding the magnesium or magnesium carbonate until . . .

_____

_____

(*a*) Complete instruction 4 of the procedure.   **1**

(*b*) Why is an excess of magnesium or magnesium carbonate added to the acid?

_____

_____   **1**

(*c*) There are three steps in the preparation of magnesium sulphate from magnesium or magnesium carbonate.

Instructions 1 to 4, shown above, describe the **"reaction step"**.

Name the next two steps.

Step 2 _____

Step 3 _____   **1**

DO NOT WRITE IN THIS MARGIN

Marks

9. (continued)

(*d*) The equation for the preparation of magnesium sulphate from magnesium is shown below.

$$Mg(s) + H_2SO_4(aq) \longrightarrow MgSO_4(aq) + H_2(g)$$

Calculate the mass of magnesium sulphate produced when 4·9 g of magnesium reacts completely with dilute sulphuric acid.

______________ g **2**

(5)

[Turn over

Marks

10. Biogas is a renewable fuel which consists of 70% methane and 30% carbon dioxide.

   (a) Name the **two** products which would be formed when biogas is burned in a good supply of air.

   _____   1

   (b) Carbohydrates in plant waste can be digested by bacteria in the soil to produce biogas.

   The table shows how the number of bacteria in soil is affected by the pH of the soil.

pH of soil	4·0	4·5	5·5	6·0	6·5
Number of bacteria (millions per gram of soil)	2·0	3·0	8·0	5·0	4·0

   (i) Draw a line graph of these results.

   (Additional graph paper, if required, can be found on page 26.)

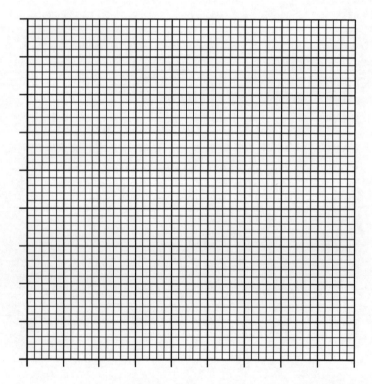

   2

   (ii) At which pH would the rate of biogas production be fastest?

   _____   1

   (4)

Marks

11. (*a*) Name the alkane shown below.

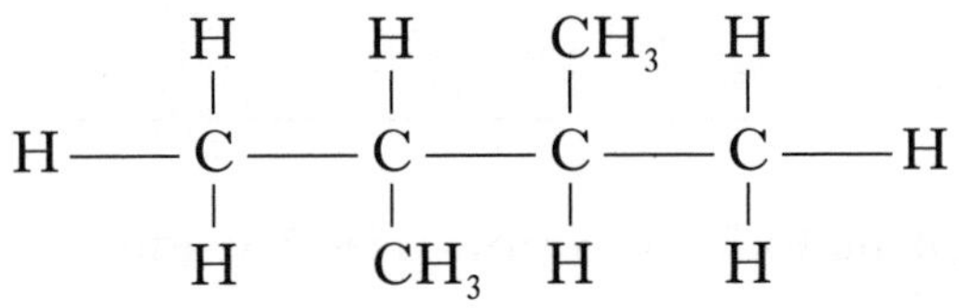

$$H-\overset{\underset{\displaystyle H}{|}}{\underset{\underset{\displaystyle H}{|}}{C}}-\overset{\underset{\displaystyle H}{|}}{\underset{\underset{\displaystyle CH_3}{|}}{C}}-\overset{\underset{\displaystyle CH_3}{|}}{\underset{\underset{\displaystyle H}{|}}{C}}-\overset{\underset{\displaystyle H}{|}}{\underset{\underset{\displaystyle H}{|}}{C}}-H$$

_____

1

(*b*) Alkanes can be reacted with alkenes to produce longer chain alkanes.

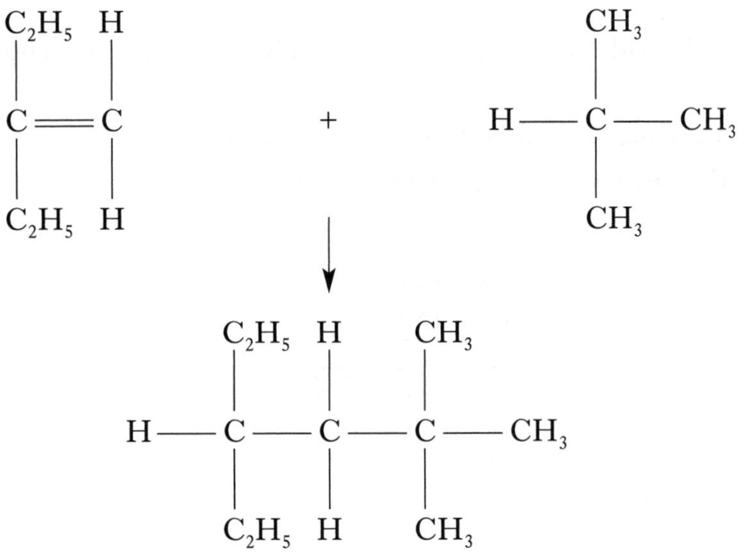

Draw the structural formula of the alkane formed in the following reaction.

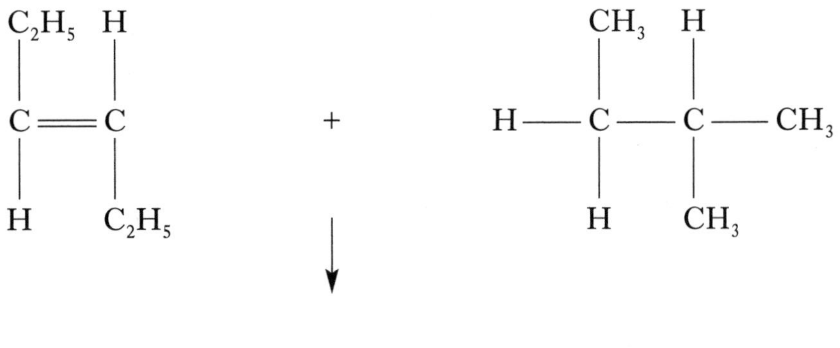

1

(2)

Marks

12. (*a*) Why does copper metal conduct electricity?

_____

_____ 1

(*b*) Copper reacts with chlorine to form copper(II) chloride.

The chloride ions in copper(II) chloride have a stable electron arrangement. How do they achieve this arrangement?

_____

_____ 1

(*c*) Copper(II) chloride solution can be electrolysed to produce copper and chlorine.

Complete the diagram below to show this electrolysis.

Show clearly where copper and chlorine are formed.

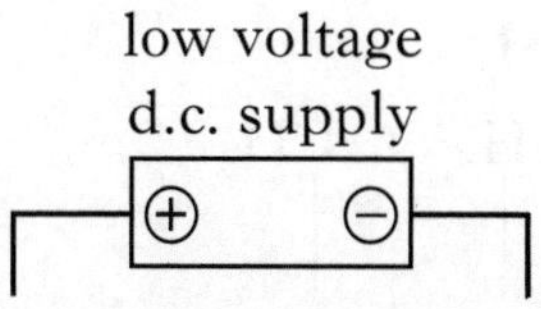

low voltage
d.c. supply

$\boxed{\oplus \qquad \ominus}$

2

(4)

Marks

13. Oil rigs are made from steel.

The oil rigs have zinc blocks attached to them to help prevent rusting. The zinc is oxidised, protecting the steel.

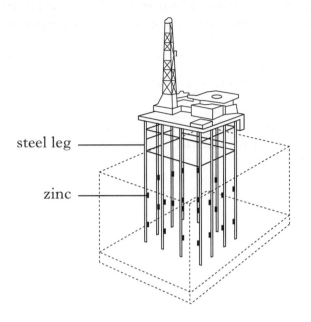

steel leg

zinc

(*a*) Write the ion-electron equation for the oxidation of zinc.

_____   **1**

(*b*) Why would tin blocks not prevent rusting?

_____

_____

_____   **1**

(*c*) Stainless steel is a type of steel which does not need protection. It contains chromium which forms an outer layer of chromium(III) oxide.

Write the formula for chromium(III) oxide.

_____   **1**

(3)

[Turn over

Marks

14. A student's results are shown below for the **PPA "Testing for Unsaturation"**.

Hydrocarbon	Molecular formula	Observation with bromine solution	Saturated or Unsaturated
A	C_6H_{14}	No change	Saturated
B	C_6H_{12}	Bromine decolourises	Unsaturated
C	C_6H_{12}	No change	Saturated
D	C_6H_{10}	Bromine decolourises	Unsaturated

(*a*) Draw a possible structural formula for hydrocarbon **D**.

1

(*b*) Sodium thiosulphate solution is made available as a safety measure, because one of the chemicals used in the experiment is corrosive.

Which chemical is corrosive?

_____

1

(2)

Marks

15. Ethers are useful chemicals.

Some ethers are listed in the table below.

Structural formula	Name of ether
$CH_3CH_2 - O - CH_2CH_3$	ethoxyethane
$CH_3 - O - CH_2CH_2CH_3$	methoxypropane
$CH_3 - O - CH_2CH_3$	methoxyethane
$CH_3CH_2 - O - CH_2CH_2CH_3$	**X**

(a) Suggest a name for ether **X**.

_____   1

(b) An ethoxyethane molecule can be formed when **two** ethanol molecules join together **with the loss of water**.

$$CH_3CH_2OH \ + \ HOCH_2CH_3 \ \rightarrow \ CH_3CH_2 - O - CH_2CH_3 \ + \ H_2O$$

Name the type of reaction taking place.

_____   1

(c) The boiling points of ethers and alkanes are approximately the same when they have a **similar** relative formula mass.

Suggest the boiling point of ethoxyethane (relative formula mass 74).

(You may wish to use page 6 of the data booklet to help you.)

_____ °C   1

(3)

[END OF QUESTION PAPER]

ADDITIONAL SPACE FOR ANSWERS

ADDITIONAL GRAPH PAPER FOR QUESTION 10(*b*)(i)

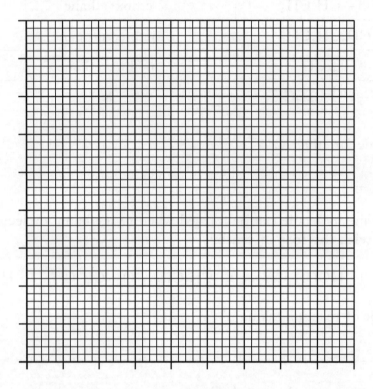

[BLANK PAGE]

FOR OFFICIAL USE

Section B   **Total Marks**

X012/201

NATIONAL
QUALIFICATIONS
2004

WEDNESDAY, 2 JUNE
9.00 AM – 11.00 AM

CHEMISTRY
INTERMEDIATE 2

Fill in these boxes and read what is printed below.

Full name of centre

Town

Forename(s)

Surname

Date of birth
Day Month Year   Scottish candidate number   Number of seat

Necessary data will be found in the Chemistry Data Booklet for Standard Grade and Intermediate 2 (1999 Edition).

Section A—Questions 1 to 30

Instructions for completion of **Section A** are given on page two.

Section B

All questions should be attempted.

The questions may be answered in any order but all answers are to be written in the spaces provided in this answer book, and must be written clearly and legibly in ink.

Rough work, if any should be necessary, as well as the fair copy, is to be written in this book.

Rough work should be scored through when the fair copy has been written.

Additional space for answers and rough work will be found at the end of the book. If further space is required, supplementary sheets may be obtained from the invigilator and should be inserted inside the **front** cover of this book.

Before leaving the examination room you must give this book to the invigilator. If you do not, you may lose all the marks for this paper.

SCOTTISH
QUALIFICATIONS
AUTHORITY

SECTION A

Read carefully

1. Check that the answer sheet provided is for Chemistry Intermediate 2 (Section A).

2. Fill in the details required on the answer sheet.

3. **In questions 1 to 30 of this part of the paper, an answer is given by indicating the choice A, B, C or D by a stroke made in INK in the appropriate place of the answer sheet—see the sample question below.**

4. **For each question there is only ONE correct answer.**

5. Rough working, if required, should be done only on this question paper, or on the rough working sheet provided—**not** on the answer sheet.

6. At the end of the examination the answer sheet for Section A **must** be placed **inside** the front cover of this answer book.

This part of the paper is worth 30 marks.

SAMPLE QUESTION

To show that the ink in a ball-pen consists of a mixture of dyes, the method of separation would be

        A fractional distillation

        B chromatography

        C fractional crystallisation

        D filtration.

The correct answer is B—chromatography. A **heavy** vertical line should be drawn joining the two dots in the appropriate box in the column headed **B** as shown **in the example on the answer sheet**.

If, after you have recorded your answer, you decide that you have made an error and wish to make a change, you should cancel the original answer and put a vertical stroke in the box you now consider to be correct. Thus, if you want to change an answer **D** to an answer **B**, your answer sheet would look like this:

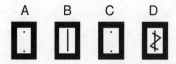

If you want to change back to an answer which has already been scored out, you should **enter a tick (✓)** to the RIGHT of the box of your choice, thus:

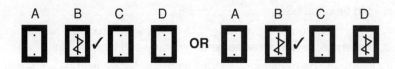

SECTION A

1. Which of the following gases is a noble gas?

   A   Nitrogen

   B   Fluorine

   C   Oxygen

   D   Neon

2. When hydrogen chloride gas is dissolved in water a solution containing hydrogen ions and chloride ions is formed.

Which equation correctly shows the state symbols for this change?

   A $HCl(g) + H_2O(\ell) \rightarrow H^+(aq) + Cl^-(aq)$

   B $HCl(\ell) + H_2O(aq) \rightarrow H^+(\ell) + Cl^-(\ell)$

   C $HCl(aq) + H_2O(\ell) \rightarrow H^+(aq) + Cl^-(aq)$

   D $HCl(g) + H_2O(\ell) \rightarrow H^+(\ell) + Cl^-(\ell)$

3. The graph below shows the variation of concentration of a reactant with time as a reaction proceeds.

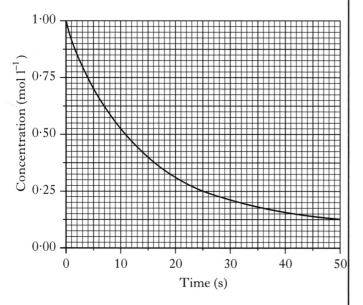

During the first 25 s, the average reaction rate, in $mol\,l^{-1}\,s^{-1}$, is

   A   0·04

   B   0·03

   C   0·02

   D   0·01.

4. The table shows the times taken for 0·5 g of magnesium to react completely with acid under different conditions.

Acid concentration $(mol\,l^{-1})$	Temperature (°C)	Reaction time (s)
0·1	20	80
0·1	25	60
0·2	30	20
0·2	40	10

The time for 0·5 g of magnesium to react completely with 0·2 $mol\,l^{-1}$ acid at 25 °C will be

   A   less than 10 s

   B   between 10 s and 20 s

   C   between 20 s and 60 s

   D   more than 80 s.

5. An atom has atomic number 26 and mass number 56.

The number of electrons in the atom is

   A   26

   B   30

   C   56

   D   82.

6. Metallic bonds are due to

   A   pairs of electrons being shared equally between atoms

   B   pairs of electrons being shared unequally between atoms

   C   the attraction of oppositely charged ions for each other

   D   the attraction of positively charged ions for delocalised electrons.

[Turn over

7. During the electrolysis of silver(I) nitrate solution, silver ions, $Ag^+(aq)$,

   A   gain electrons at the negative electrode

   B   lose electrons at the negative electrode

   C   gain electrons at the positive electrode

   D   lose electrons at the positive electrode.

8. Which of the following could be the molecular formula of a cycloalkane?

   A C_7H_{16}

   B C_7H_{14}

   C C_7H_{12}

   D C_7H_{10}

9. Part of a polymer structure is shown.

$$\begin{array}{cccccc} CH_3 & H & CH_3 & H & CH_3 & H \\ | & | & | & | & | & | \\ -C & -C & -C & -C & -C & -C- \\ | & | & | & | & | & | \\ H & CN & H & CN & H & CN \end{array}$$

Which of the following gases could **not** be produced when this polymer is burned?

   A CO

   B CO_2

   C HCl

   D HCN

10. The conversion of an oil into a hardened fat involves the

   A   addition of water

   B   removal of water

   C   addition of hydrogen

   D   removal of hydrogen.

11.

$$\begin{array}{ccccc} H & H & H & CH_3 & H \\ | & | & | & | & | \\ H-C & -C & -C & -C & -C-H \\ | & | & | & | & | \\ H & H & CH_3 & H & H \end{array}$$

The name of the above compound is

   A   2,3-dimethylpentane

   B   3,4-dimethylpentane

   C   2,3-dimethylpropane

   D   3,4-dimethylpropane.

12. Which of the following compounds belongs to the same homologous series as the compound with the molecular formula C_3H_8?

A

$$\begin{array}{cc} H & H \\ | & | \\ H-C & -C-H \\ | & | \\ H-C & -C-H \\ | & | \\ H & H \end{array}$$

B

$$\begin{array}{cccc} H & & & H \\ | & & & | \\ H-C & -C & =C & -C-H \\ | & | & | & | \\ H & H & H & H \end{array}$$

C

$$\begin{array}{cccc} & & H & \\ & & | & \\ & & H-C-H & \\ H & H & | & H \\ | & | & | & | \\ H-C & -C & -C & -C-H \\ | & | & | & | \\ H & H & H & H \end{array}$$

D

$$\begin{array}{cccc} & & H & \\ & & | & \\ & & H-C-H & \\ H & H & | & \\ | & | & | & \\ H-C & -C & -C & =C-H \\ | & | & & | \\ H & H & & H \end{array}$$

13. The first four members of the alkanal homologous series are:

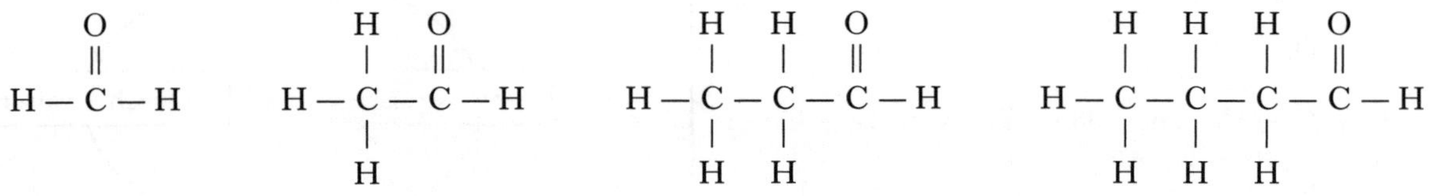

$$
\begin{array}{cccc}
\underset{\displaystyle H-\overset{\displaystyle O}{\overset{\|}{C}}-H}{} &
\underset{\displaystyle H-\overset{\overset{\displaystyle H}{|}}{\underset{\underset{\displaystyle H}{|}}{C}}-\overset{\displaystyle O}{\overset{\|}{C}}-H}{} &
\underset{\displaystyle H-\overset{\overset{\displaystyle H}{|}}{\underset{\underset{\displaystyle H}{|}}{C}}-\overset{\overset{\displaystyle H}{|}}{\underset{\underset{\displaystyle H}{|}}{C}}-\overset{\displaystyle O}{\overset{\|}{C}}-H}{} &
\underset{\displaystyle H-\overset{\overset{\displaystyle H}{|}}{\underset{\underset{\displaystyle H}{|}}{C}}-\overset{\overset{\displaystyle H}{|}}{\underset{\underset{\displaystyle H}{|}}{C}}-\overset{\overset{\displaystyle H}{|}}{\underset{\underset{\displaystyle H}{|}}{C}}-\overset{\displaystyle O}{\overset{\|}{C}}-H}{}
\end{array}
$$

The general formula for this homologous series is

A $C_nH_{2n-2}O$

B $C_nH_{2n}O$

C $C_nH_{2n+1}O$

D $C_nH_{2n+2}O.$

14. Industrial ethanol can be manufactured from ethene by

A   condensation

B   dehydration

C   hydration

D   hydrolysis.

15. The structure below shows a section of an addition polymer.

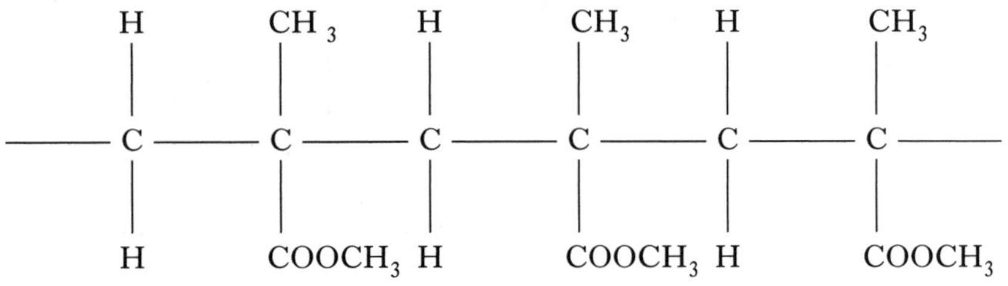

$$
-\overset{\overset{\displaystyle H}{|}}{\underset{\underset{\displaystyle H}{|}}{C}}-\overset{\overset{\displaystyle CH_3}{|}}{\underset{\underset{\displaystyle COOCH_3}{|}}{C}}-\overset{\overset{\displaystyle H}{|}}{\underset{\underset{\displaystyle H}{|}}{C}}-\overset{\overset{\displaystyle CH_3}{|}}{\underset{\underset{\displaystyle COOCH_3}{|}}{C}}-\overset{\overset{\displaystyle H}{|}}{\underset{\underset{\displaystyle H}{|}}{C}}-\overset{\overset{\displaystyle CH_3}{|}}{\underset{\underset{\displaystyle COOCH_3}{|}}{C}}-
$$

Which molecule is used to make this polymer?

A $\overset{\overset{\displaystyle CH_3}{|}}{\underset{\underset{\displaystyle H}{|}}{C}}=\overset{\overset{\displaystyle H}{|}}{\underset{\underset{\displaystyle COOCH_3}{|}}{C}}$

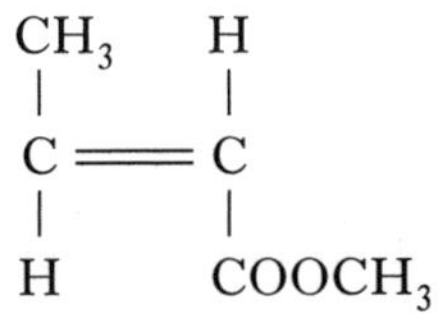

B $\overset{\overset{\displaystyle H}{|}}{\underset{\underset{\displaystyle H}{|}}{C}}=\overset{\overset{\displaystyle CH_3}{|}}{\underset{\underset{\displaystyle COOCH_3}{|}}{C}}$

C $\overset{\overset{\displaystyle CH_3}{|}}{\underset{\underset{\displaystyle H}{|}}{C}}=\overset{\overset{\displaystyle COOCH_3}{|}}{\underset{\underset{\displaystyle H}{|}}{C}}$

D $H-\overset{\overset{\displaystyle H}{|}}{\underset{\underset{\displaystyle H}{|}}{C}}-\overset{\overset{\displaystyle CH_3}{|}}{\underset{\underset{\displaystyle COOCH_3}{|}}{C}}-H$

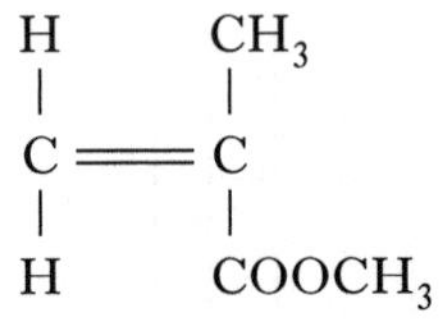

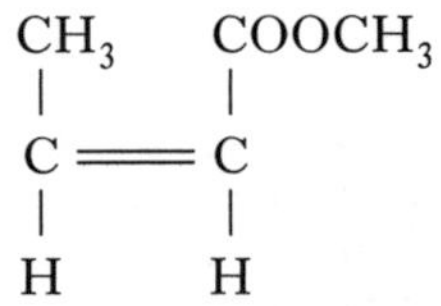

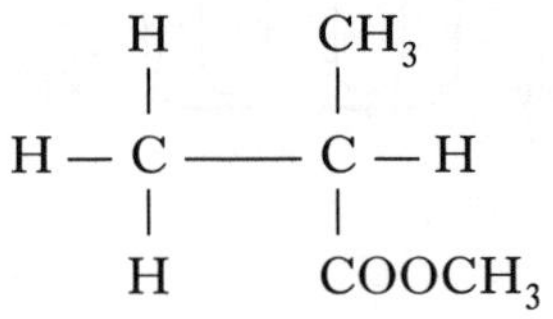

16. Two isomers of butene are

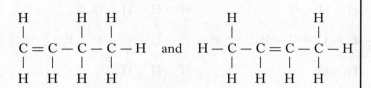

Which of the following structures represents a third isomer of butene?

A

$$H - C - C - C = C$$

B

C

D

17.

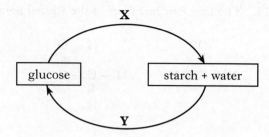

Which line in the table correctly describes reactions **X** and **Y**?

	Reaction X	**Reaction Y**
A	condensation	hydration
B	condensation	hydrolysis
C	dehydration	hydration
D	dehydration	hydrolysis

18. In which of the following experiments would both carbohydrates give an orange precipitate when heated with Benedict's solution?

A

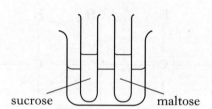

sucrose   maltose

B

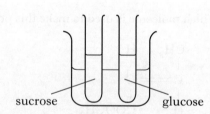

sucrose   glucose

C

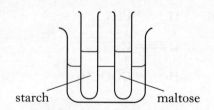

starch   maltose

D

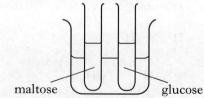

maltose   glucose

19. Which structure represents a molecule of glycerol?

A

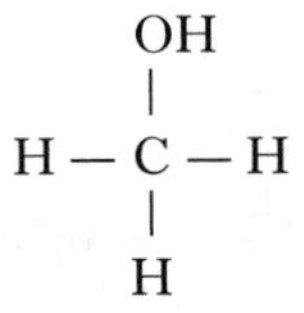

$$H - \overset{\overset{\displaystyle OH}{|}}{\underset{\underset{\displaystyle H}{|}}{C}} - H$$

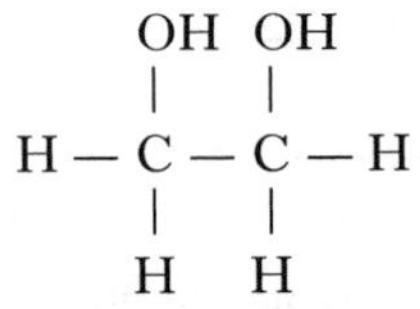

B

$$H - \overset{\overset{\displaystyle OH}{|}}{\underset{\underset{\displaystyle H}{|}}{C}} - \overset{\overset{\displaystyle OH}{|}}{\underset{\underset{\displaystyle H}{|}}{C}} - H$$

C

$$H - \overset{\overset{\displaystyle OH}{|}}{\underset{\underset{\displaystyle H}{|}}{C}} - \overset{\overset{\displaystyle OH}{|}}{\underset{\underset{\displaystyle H}{|}}{C}} - \overset{\overset{\displaystyle OH}{|}}{\underset{\underset{\displaystyle H}{|}}{C}} - H$$

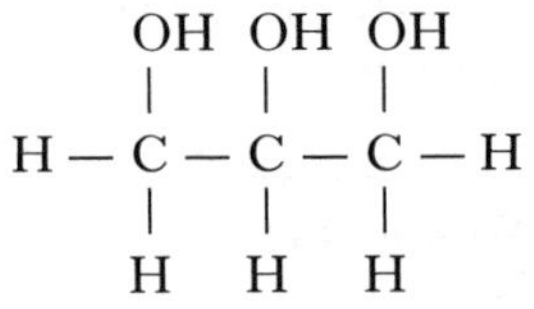

D

$$H - \overset{\overset{\displaystyle OH}{|}}{\underset{\underset{\displaystyle H}{|}}{C}} - \overset{\overset{\displaystyle OH}{|}}{\underset{\underset{\displaystyle H}{|}}{C}} - \overset{\overset{\displaystyle OH}{|}}{\underset{\underset{\displaystyle H}{|}}{C}} - \overset{\overset{\displaystyle OH}{|}}{\underset{\underset{\displaystyle H}{|}}{C}} - H$$

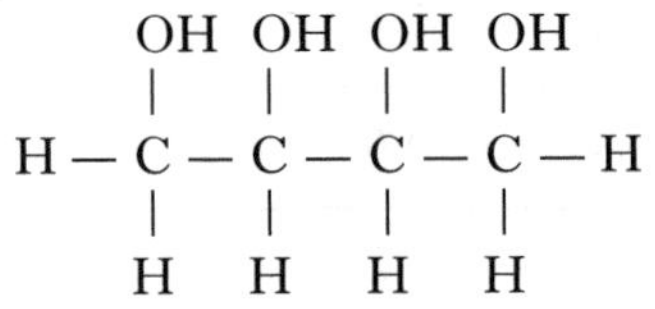

20. Which type of compound is represented by the structure shown?

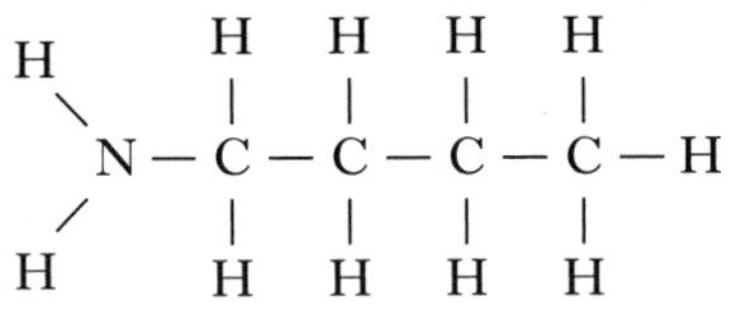

$$\overset{H}{\underset{H}{>}}N - \overset{\overset{\displaystyle H}{|}}{\underset{\underset{\displaystyle H}{|}}{C}} - \overset{\overset{\displaystyle H}{|}}{\underset{\underset{\displaystyle H}{|}}{C}} - \overset{\overset{\displaystyle H}{|}}{\underset{\underset{\displaystyle H}{|}}{C}} - \overset{\overset{\displaystyle H}{|}}{\underset{\underset{\displaystyle H}{|}}{C}} - H$$

A   Amine

B   Protein

C   Amino acid

D   Carboxylic acid

21. 0·5 mol of pure citric acid was dissolved in water and the solution made up to 250 cm³.

What was the concentration of the solution formed?

A $0{\cdot}25 \text{ mol } l^{-1}$

B $0{\cdot}5 \text{ mol } l^{-1}$

C $1{\cdot}0 \text{ mol } l^{-1}$

D $2{\cdot}0 \text{ mol } l^{-1}$

22. $Fe^{3+}(aq) + e^- \rightarrow Fe^{2+}(aq)$

This ion-electron equation represents

A   oxidation of iron(III) ions

B   oxidation of iron(II) ions

C   reduction of iron(III) ions

D   reduction of iron(II) ions.

23. Which of the following oxides, when shaken with water, would leave the pH unchanged?

(You may wish to use page 5 of the data booklet to help you.)

A   Carbon dioxide

B   Copper oxide

C   Sodium oxide

D   Sulphur dioxide

24. Which of the following compounds is a base?

A   Magnesium carbonate

B   Magnesium chloride

C   Magnesium nitrate

D   Magnesium sulphate

25. Which of the following solutions, when added to copper chloride solution, produces a precipitate?

(You may wish to use page 5 of the data booklet to help you.)

A   Calcium bromide solution

B   Lithium sulphate solution

C   Magnesium nitrate solution

D   Sodium hydroxide solution

26. Which of the following gases reacts with an alkaline solution?

A   Nitrogen dioxide

B   Ammonia

C   Oxygen

D   Argon

[Turn over

27.

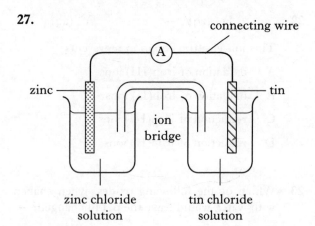

connecting wire

zinc — tin

ion bridge

zinc chloride solution — tin chloride solution

In the cell shown above, electrons flow through

A   the solution from tin to zinc

B   the solution from zinc to tin

C   the connecting wire from tin to zinc

D   the connecting wire from zinc to tin.

28. Which of the following solutions will react with iron metal?

(You may wish to use page 7 of the data booklet to help you.)

A   Magnesium chloride

B   Tin chloride

C   Sodium chloride

D   Zinc chloride

29. Four cells were made by joining copper, iron, silver and tin to zinc.

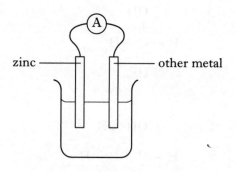

zinc — other metal

Which line in the table shows the voltage of the cell containing iron joined to zinc?

(You may wish to use page 7 of the data booklet to help you.)

Cell	Voltage (V)
A	1·5
B	1·1
C	0·6
D	0·3

30. Which of the following metals must be obtained from its ore by electrolysis?

A   Aluminium

B   Copper

C   Iron

D   Gold

Candidates are reminded that the answer sheet for Section A MUST be placed INSIDE the front cover of this answer book.

Turn over for SECTION B on *Page ten*

DO NOT
WRITE
THIS
MARGIN

Marks

SECTION B

50 marks are available in this section of the paper.

1. The nuclide notation for an isotope of hydrogen is $_1^1$H.

   (a) An isotope of bromine has atomic number 35 and mass number 81.

      (i) Complete the nuclide notation for this isotope of bromine.

   Br

   1

      (ii) How many neutrons are there in this isotope?

   _____

   1

   (b) Bromine has two isotopes. One has a mass number of 81 and the other has a mass number of 79.

   The relative atomic mass of bromine is 80.

   What does this tell you about the percentage of each isotope in bromine?

   _____

   _____   1

   (3)

Marks

2. In a catalytic converter, harmful gases are changed to less harmful gases by passing them over a solid catalyst as shown in the diagram.

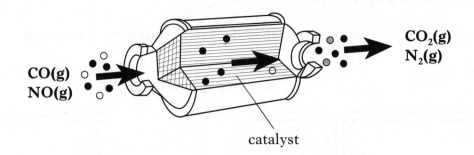

catalyst

(a) Balance the equation for this reaction.

$$CO(g) \quad + \quad NO(g) \quad \longrightarrow \quad CO_2(g) \quad + \quad N_2(g)$$

1

(b) Why can the catalyst be described as a heterogeneous catalyst?

_____

_____ **1**

(2)

[Turn over

Marks

3. Candle wax is a hydrocarbon.

Blue cobalt chloride paper and limewater can be used to detect products formed when candle wax is burned.

(a) Complete and label the diagram to show the arrangement you would use.

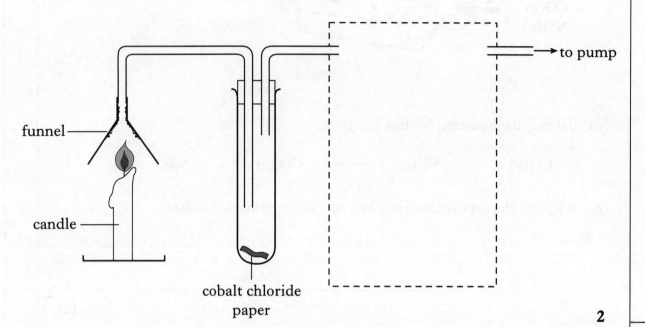

funnel

candle

cobalt chloride
paper

to pump

2

(b) Name the product detected using the blue cobalt chloride paper.

_____

1

(c) As the candle burns the funnel becomes coated with soot.
Why does this happen?

_____

_____

1

(4)

DO NOT WRITE IN THIS MARGIN

Marks

4. Hydrogen reacts with chlorine to form hydrogen chloride.

(*a*) Draw a diagram to show how the **outer** electrons are shared in a molecule of hydrogen chloride. **1**

(*b*) Chlorine has a greater attraction than hydrogen for the bonded electrons in a hydrogen chloride molecule.

What term is used to describe this type of covalent bond?

____ **1**

(2)

[Turn over

Marks

5. Urea is a chemical which is present in urine.

   (a) Urea has the formula **H₂NCONH₂**.
       Draw the full structural formula for urea.

1

   (b) An adult male, on average, excretes 30 g of urea each day.
       How many moles are there in 30 g of urea?

1

   (c) Urea can be hydrolysed to form carbon dioxide and ammonia.

   $$H_2NCONH_2(aq) \ + \ H_2O(\ell) \ \longrightarrow \ CO_2(aq) \ + \ 2NH_3(aq)$$

   This reaction was carried out with different concentrations of urea and
   the relative rates of hydrolysis were calculated.

   Results

Concentration of urea (mol l⁻¹)	Relative rate
0·05	4·4
0·10	5·9
0·30	7·7
0·40	8·0
0·50	8·1

Marks

5. **(c)** **(continued)**

(i) Plot these results as a line graph.

Use appropriate scales to fill most of the graph paper.

(Additional graph paper, if required, can be found on page 27.)

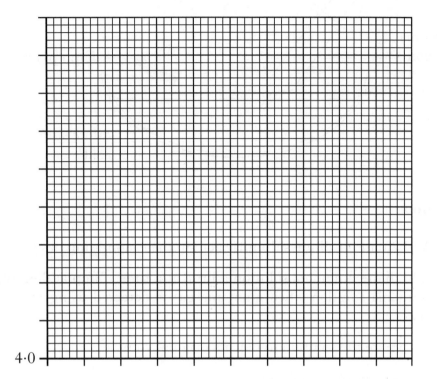

4·0

2

(ii) Using your graph, estimate the concentration of urea which gives a relative rate of 7·2.

_____ mol l^{-1} 1

(5)

[Turn over

DO NOT
WRITE I
THIS
MARGI

Marks

6.

"Making Ammonium Sulphate in the lab"

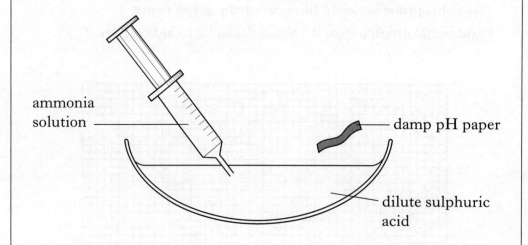

ammonia solution ——— damp pH paper

dilute sulphuric acid

Method

1. Add 20 cm³ of dilute sulphuric acid to the evaporating basin.

2. Add dilute ammonia solution a little at a time. Stop adding the ammonia solution when ammonia gas is detected using the pH paper.

3. Evaporate the solution to half its original volume.

4. Set aside until crystals form.

(*a*) The chemical equation for the reaction is

$$2NH_4^+(aq) + 2OH^-(aq) + 2H^+(aq) + SO_4^{2-}(aq) \rightarrow 2NH_4^+(aq) + SO_4^{2-}(aq) + 2H_2O(\ell)$$

Rewrite the equation omitting the spectator ions.

_____ 1

(*b*) How will ammonia gas affect damp pH paper?

_____

_____ 1

(*c*) State a use for ammonium sulphate.

_____ 1

(3)

DO NOT WRITE IN THIS MARGIN

Marks

7. The diagram below shows the apparatus used in the **PPA, "Cracking"**.

Liquid paraffin is cracked using an aluminium oxide catalyst.

Bromine solution is used to show that some of the products are unsaturated.

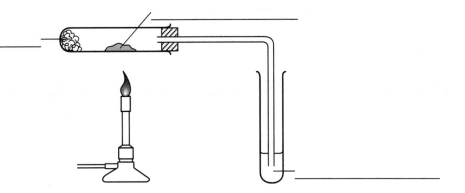

(*a*) Label the diagram of the apparatus used to crack liquid paraffin. **1**

(*b*) What safety precaution should be taken before heating is stopped?

____

____ **1**

(2)

[Turn over

Marks

8. Lawn fertiliser can contain iron(II) sulphate to kill moss in the lawn.

   (a) Iron(II) sulphate can be made by reacting iron(II) oxide with dilute sulphuric acid.

$$FeO(s) \ + \ H_2SO_4(aq) \ \longrightarrow \ FeSO_4(aq) \ + \ H_2O(\ell)$$

   Name this type of chemical reaction.

   _____ 1

   (b) What mass of iron(II) sulphate can be made from 144 kg of iron(II) oxide?

   _____ kg

   2
   (3)

Marks

9.   **"A method of making sulphuric acid"**

Oxidise ammonia to produce nitrogen monoxide.   Water will also be
produced in the reaction.   React the nitrogen monoxide with more oxygen
to form nitrogen dioxide.   The sulphuric acid can then be produced by
reacting the nitrogen dioxide with sulphurous acid.

(*a*) Complete the flow chart to show this process.

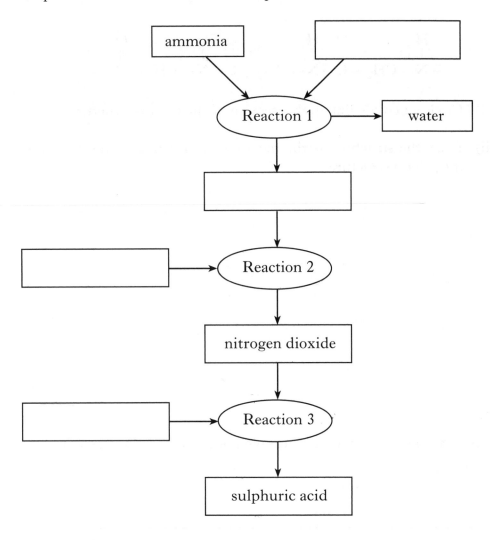

2

(*b*) When nitrogen dioxide reacts with sulphurous acid to produce
sulphuric acid, nitrogen monoxide will also be produced. This can be
recycled.

Add a line to the flow chart to show nitrogen monoxide being recycled.   1

(*c*) Sulphuric acid is a strong acid and sulphurous acid is a weak acid.

Why is sulphuric acid described as a strong acid?

_____

_____   1

(4)

[Turn over

Marks

10. Jelly is made from a protein called gelatin.

(a) Name the type of monomers which join together to form proteins.

_____

1

(b) A section of the gelatin structure is shown below.

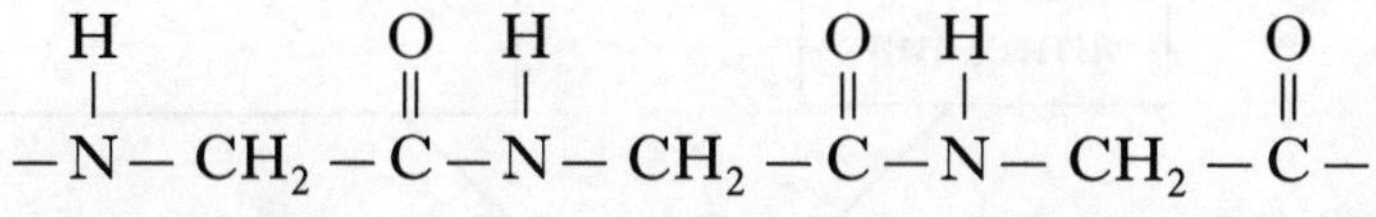

$$\begin{array}{ccccccccc} & H & & O & H & & O & H & & O \\ & | & & \parallel & | & & \parallel & | & & \parallel \\ -N- & CH_2 & -C- & N- & CH_2 & -C- & N- & CH_2 & -C- \end{array}$$

(i) Circle a peptide link in this section of the gelatin structure.

1

(ii) Draw the structure of the monomer which makes this section of the gelatin structure.

1

(c) Papain, an enzyme found in pineapple juice, can hydrolyse gelatin.

When papain is heated to a high temperature it no longer hydrolyses the gelatin. Suggest why.

_____

_____

1

(4)

Marks

11. Ethyl ethanoate can be made by reacting ethanoic acid and ethanol.

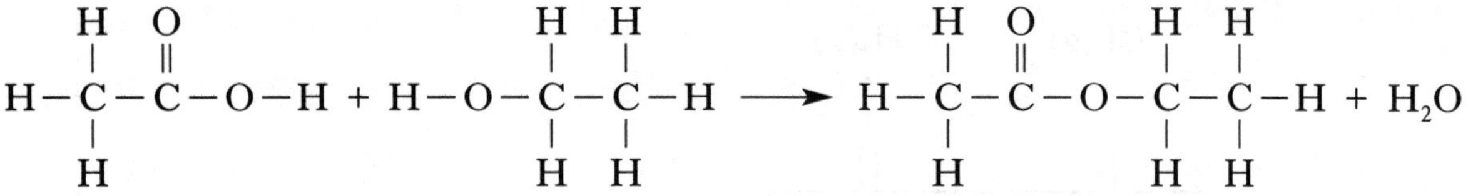

$$\begin{array}{c} H \quad O \\ | \quad \| \\ H-C-C-O-H \\ | \\ H \end{array} + \begin{array}{c} H \quad H \\ | \quad | \\ H-O-C-C-H \\ | \quad | \\ H \quad H \end{array} \longrightarrow \begin{array}{c} H \quad O \quad\quad H \quad H \\ | \quad \| \quad\quad | \quad | \\ H-C-C-O-C-C-H \\ | \quad\quad\quad | \quad | \\ H \quad\quad\quad H \quad H \end{array} + H_2O$$

(a) Name this type of chemical reaction.

_____   1

(b) A method of making ethyl ethanoate from ethanol **only**, has been developed.

$$2C_2H_5OH \xrightarrow{\text{catalyst}} CH_3COOC_2H_5 + 2X$$

   (i) Name substance **X**.

_____   1

   (ii) This method was developed for use in countries where ethanol is made from a renewable source.

Name this source of ethanol.

_____   1

                                                (3)

[Turn over

Marks

12. A membrane cell can be used to produce **chlorine, hydrogen** and **sodium hydroxide** from sodium chloride solution.

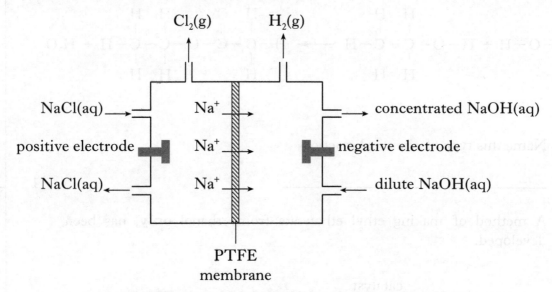

Electrode	Ion-electron equation
positive	$2Cl^- \longrightarrow Cl_2 + 2e^-$
negative	$2H_2O + 2e^- \longrightarrow 2OH^- + H_2$

(a) Combine the ion-electron equations to produce a redox equation.

_____   1

(b) The membrane is made up of layers of poly(tetrafluoroethene) (PTFE) with negatively charged groups attached. Sodium ions pass through the membrane towards the negative electrode.

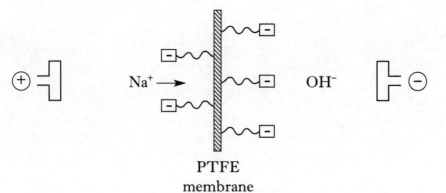

Suggest a reason why hydroxide ions (OH^-) will not pass through the membrane towards the positive electrode.

_____   1

(c) PTFE can be re-shaped on heating.

What name is given to plastics which can be re-shaped on heating?

_____   1

DO NOT WRITE IN THIS MARGIN

Marks

13. (*a*) The explosive, TNT, is a covalent compound with a low melting point.

Which type of covalent structure does TNT have?

____ **1**

(*b*) The equation shows the reaction when the compound, TNT, explodes.

TNT + oxygen ⟶ carbon dioxide + nitrogen + water

Using the equation, name the elements that **must** be present in TNT.

____ **1**

(2)

[Turn over

Marks

14. In the **Unit 1 PPA, "Effect of concentration on reaction rate"**, the reaction between sodium persulphate and potassium iodide solutions is studied.

Set-up for Experiment 1

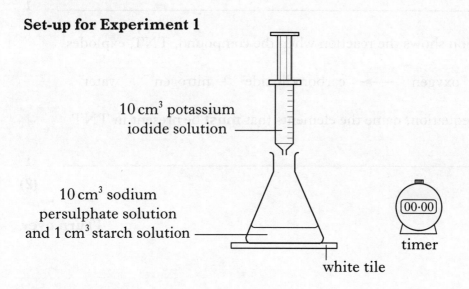

10 cm³ potassium iodide solution

10 cm³ sodium persulphate solution and 1 cm³ starch solution

white tile

timer

00·00

Results

Experiment	1	2	3	4
Volume of sodium persulphate solution (cm³)	10	8	6	4
Volume of water (cm³)	0			
Reaction time (min)	2·1	2·7	3·5	5·6

(a) Complete the table to show the volumes of water used in experiments 2, 3 and 4.

1

(b) How would you know when to stop the timer in each experiment?

_____

_____ 1

(c) Why are volumes of sodium persulphate solution less than 4 cm³ not used?

_____

_____ 1

(d) The formula for the persulphate ion is $S_2O_8^{2-}$.

Write the formula for sodium persulphate.

_____ 1

(4)

Marks

15. Iron is used to make ships.

Sometimes when a ship docks in a harbour, it is connected to the negative terminal of a power supply. A lump of scrap iron is connected to the positive terminal.

This is used to prevent the iron in the ship from rusting.

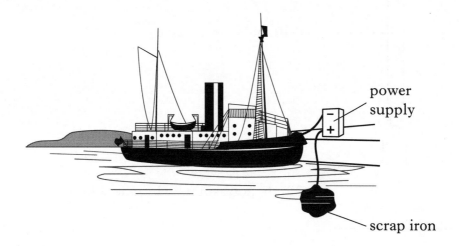

power supply

scrap iron

(*a*) (i) Why does connecting the ship to the negative terminal of the power supply prevent the iron in the ship from rusting?

_____

_____ 1

(ii) Why is sea water able to complete the circuit?

_____

_____ 1

(*b*) Suggest another method which could be used to protect the iron in the ship from rusting?

_____

_____ 1

(3)

[Turn over for Question 16 on *Page twenty-six*

Marks

16. A student carried out a titration to find the concentration of a potassium hydroxide solution using $0 \cdot 10$ mol l^{-1} nitric acid.

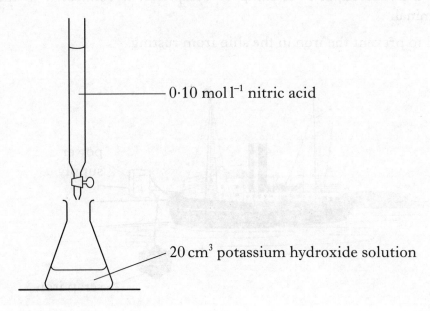

$0 \cdot 10$ mol l^{-1} nitric acid

20 cm^3 potassium hydroxide solution

Equation

$$KOH(aq) \; + \; HNO_3(aq) \; \longrightarrow \; KNO_3(aq) \; + \; H_2O(\ell)$$

(a) What must be added to the flask to show the end-point of the titration?

_____

1

(b) The average volume of nitric acid needed to neutralise the potassium hydroxide solution is $24 \cdot 6$ cm^3.

Calculate the concentration of the potassium hydroxide solution.

_____ mol l^{-1}

2

(3)

[END OF QUESTION PAPER]

ADDITIONAL SPACE FOR ANSWERS

ADDITIONAL GRAPH PAPER FOR QUESTION 5(*c*)(i)

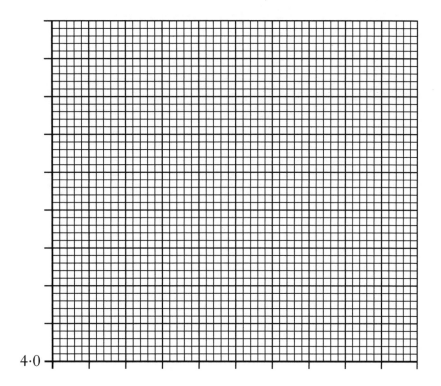

4·0

Page twenty-seven **[Turn over**

ADDITIONAL SPACE FOR ANSWERS

[BLANK PAGE]

FOR OFFICIAL USE

Section B **Total Marks**

X012/201

NATIONAL
QUALIFICATIONS
2005

TUESDAY, 31 MAY
9.00 AM – 11.00 AM

CHEMISTRY
INTERMEDIATE 2

Fill in these boxes and read what is printed below.

Full name of centre

Town

Forename(s)

Surname

Date of birth
Day Month Year   Scottish candidate number   Number of seat

Necessary data will be found in the Chemistry Data Booklet for Standard Grade and Intermediate 2 (1999 Edition).

Section A – Questions 1–30 (30 marks)

Instructions for completion of **Section A** are given on page two.

Section B (50 marks)

All questions should be attempted.

The questions may be answered in any order but all answers are to be written in the spaces provided in this answer book, and must be written clearly and legibly in ink.

Rough work, if any should be necessary, should be written in this book, and then scored through when the fair copy has been written. If further space is required, a supplementary sheet for rough work may be obtained from the invigilator.

Additional space for answers will be found at the end of the book. If further space is required, supplementary sheets may be obtained from the invigilator and should be inserted inside the **front** cover of this booklet.

Before leaving the examination room you must give this book to the invigilator. If you do not, you may lose all the marks for this paper.

SCOTTISH
QUALIFICATIONS
AUTHORITY

©

Read carefully

1   Check that the answer sheet provided is for **Chemistry Intermediate 2 (Section A)**.

2   Check that the answer sheet you have been given has **your name**, **date of birth**, **SCN** (Scottish Candidate Number) and **Centre Name** printed on it.

Do not change any of these details.

3   If any of this information is wrong, tell the Invigilator immediately.

4   If this information is correct, **print** your name and seat number in the boxes provided.

5   Use **black** or **blue ink** for your answers. **Do not use red ink**.

6   The answer to each question is **either** A, B, C or D. Decide what your answer is, then put a horizontal line in the space provided (see sample question below).

7   There is **only one correct** answer to each question.

8   Any rough working should be done on the question paper or the rough working sheet, **not** on your answer sheet.

9   At the end of the exam, put the **answer sheet for Section A inside the front cover of this answer book**.

Sample Question

To show that the ink in a ball-pen consists of a mixture of dyes, the method of separation would be

> A   fractional distillation
>
> B   chromatography
>
> C   fractional crystallisation
>
> D   filtration.

The correct answer is **B**—chromatography. The answer **B** has been clearly marked with a horizontal line (see below).

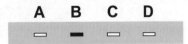

Changing an answer

If you decide to change your answer, cancel your first answer by putting a cross through it (see below) and fill in the answer you want. The answer below has been changed to **B**.

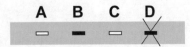

If you then decide to change back to an answer you have already scored out, put a tick (✓) to the **right** of the answer you want, as shown below:

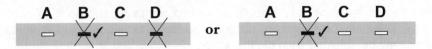

SECTION A

1. Graph P shows the volume of hydrogen gas collected when $1.0\,g$ of magnesium ribbon reacts with excess $2\ mol\,l^{-1}$ hydrochloric acid.

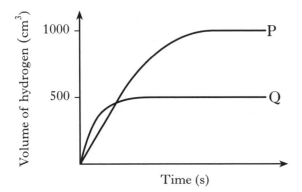

   Which of the following samples of magnesium, when reacted with excess $2\ mol\,l^{-1}$ hydrochloric acid would produce graph Q.

   A $0.5\,g$ of magnesium ribbon

   B $0.5\,g$ of magnesium powder

   C $1.0\,g$ of magnesium powder

   D $2.0\,g$ of magnesium ribbon

2. Which of the following compounds contains only two elements?

   A   Magnesium hydroxide

   B   Magnesium phosphate

   C   Magnesium sulphite

   D   Magnesium nitride

3. Which of the following gases is unchanged when it passes through the catalytic converter in a car?

   A   Carbon dioxide

   B   Carbon monoxide

   C   Nitrogen dioxide

   D   Nitrogen monoxide

4. 2, 8, 8 is the electron arrangement for an atom of an element belonging to the

   A   halogens

   B   noble gases

   C   alkali metals

   D   transition metals.

5. Different atoms of the same element have identical

   A   nuclei

   B   mass numbers

   C   numbers of neutrons

   D   numbers of protons.

6. Which of the following compounds exists as diatomic molecules?

   A   Carbon monoxide

   B   Carbon tetrachloride

   C   Nitrogen trihydride

   D   Sulphur dioxide

7. A substance, **X**, has a melting point of $996\,°C$ and a boiling point of $1704\,°C$. It only conducts electricity when molten or when dissolved in water.

   The structure of **X** is likely to be

   A   ionic

   B   metallic

   C   covalent network

   D   covalent molecular.

[Turn over

8. In a reaction, calcium silicate is decomposed to calcium oxide and silicon dioxide.

$$CaSiO_3(s) \rightarrow CaO(s) + SiO_2(s)$$

Which of the following statements about the reaction is correct?

A   The number of moles of products is less than the number of moles of reactant.

B   The number of moles of products equals the number of moles of reactant.

C   The mass of the products is greater than the mass of the reactant.

D   The mass of the products equals the mass of the reactant.

9. What name is given to the reaction shown by the following equation?

$$C_6H_{12}O_6 + 6O_2 \rightarrow 6CO_2 + 6H_2O$$

A   Combustion

B   Condensation

C   Dehydration

D   Hydrolysis

10.

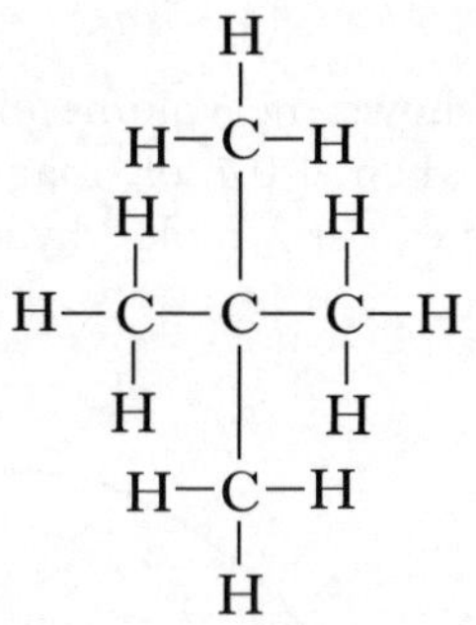

Which of the following compounds is an isomer of the one shown above?

A

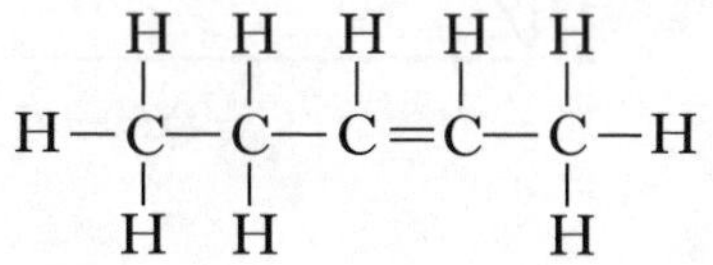

B

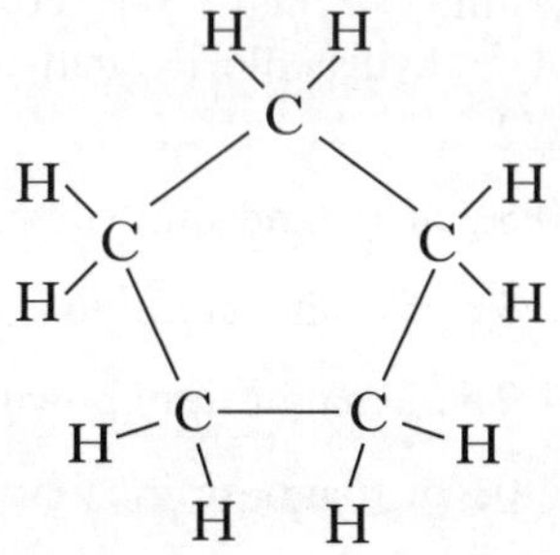

C

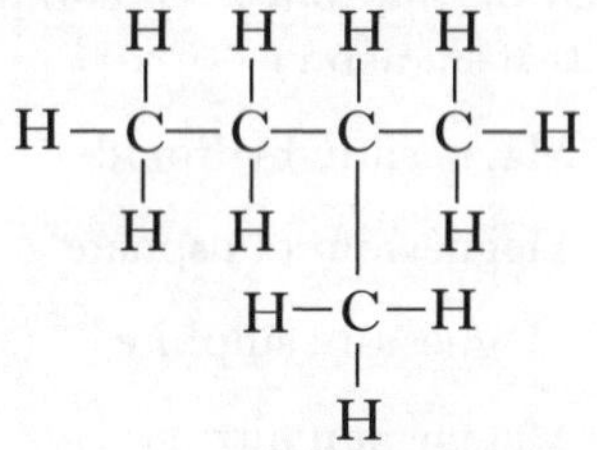

D

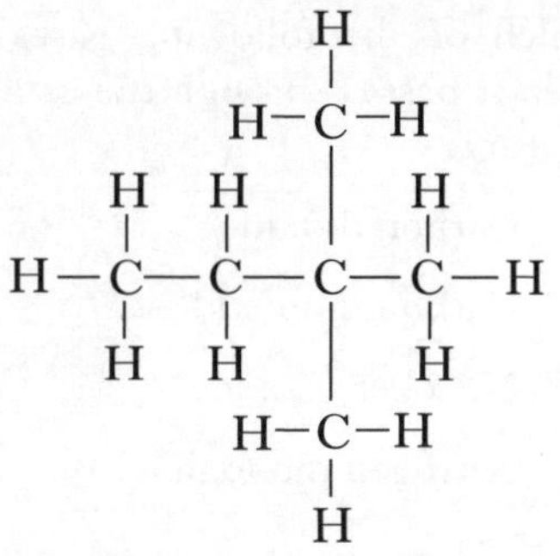

11.

$$H-\underset{\underset{H}{|}}{\overset{\overset{H}{|}}{C}}-\underset{}{\overset{\overset{CH_3}{|}}{C}}=\underset{}{\overset{\overset{H}{|}}{C}}-\underset{\underset{H}{|}}{\overset{\overset{H}{|}}{C}}-O-H$$

The above molecule is an example of

A   a saturated alcohol

B   an unsaturated alcohol

C   a saturated carboxylic acid

D   an unsaturated carboxylic acid.

12.

$$H-\underset{\underset{H}{|}}{\overset{\overset{H}{|}}{C}}-\underset{\underset{H}{|}}{\overset{\overset{H}{|}}{C}}-\underset{\underset{H}{|}}{C}=\underset{\underset{H}{|}}{C}-\underset{\underset{H}{|}}{\overset{\overset{H}{|}}{C}}-H$$

The name of the above compound is

A   but-2-ene

B   pent-2-ene

C   but-3-ene

D   pent-3-ene.

13. Which of the following structural formulae is that of an ester?

A

$$H-\underset{\underset{H}{|}}{\overset{\overset{H}{|}}{C}}-\underset{\underset{H}{|}}{\overset{\overset{H}{|}}{C}}-OH$$

B

$$H-\underset{\underset{H}{|}}{\overset{\overset{H}{|}}{C}}-C\overset{\nearrow O}{\underset{\searrow H}{}}$$

C

$$H-\underset{\underset{H}{|}}{\overset{\overset{H}{|}}{C}}-C\overset{\nearrow O}{\underset{\searrow OH}{}}$$

D

$$H-C\overset{\nearrow O}{\underset{\searrow O-\underset{\underset{H}{|}}{\overset{\overset{H}{|}}{C}}-H}{}}$$

14. A hydrocarbon **Y** has molecular formula C_6H_{12} and does not react with bromine solution.

A possible full structural formula for hydrocarbon **Y** is

A

$$H-\underset{\underset{H}{|}}{\overset{\overset{H}{|}}{C}}-\underset{\underset{H}{|}}{\overset{\overset{H-C-H}{|}}{C}}-\underset{\underset{H-C-H}{|}}{\overset{\overset{H}{|}}{C}}-\underset{\underset{H}{|}}{\overset{\overset{H}{|}}{C}}-H$$

B

(cyclohexane ring structure)

C

$$\underset{\underset{H}{|}}{\overset{\overset{H}{|}}{C}}=\underset{\underset{H}{|}}{\overset{\overset{H}{|}}{C}}-\underset{\underset{H}{|}}{\overset{\overset{H}{|}}{C}}-\underset{\underset{H}{|}}{\overset{\overset{H}{|}}{C}}-\underset{\underset{H}{|}}{\overset{\overset{H}{|}}{C}}-\underset{\underset{H}{|}}{\overset{\overset{H}{|}}{C}}-H$$

D

(cyclohexene ring structure)

[Turn over

15. Which of the following is a renewable source of energy?

    A   Coal

    B   Petrol

    C   Ethanol

    D   Natural gas

16.

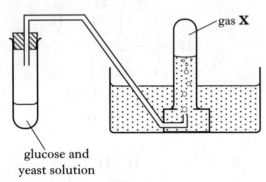

glucose and
yeast solution

Gas **X** will

    A   burn with a pop

    B   turn limewater cloudy

    C   relight a glowing splint

    D   rapidly decolourise bromine solution.

17. Which of the following is the molecular formula for a carbohydrate?

    A $C_6H_{14}O$

    B $C_6H_{12}O_2$

    C $C_6H_{10}O_4$

    D $C_6H_{12}O_6$

18. What type of substance is formed when a protein is hydrolysed?

    A   A sugar

    B   An alkanol

    C   An amino acid

    D   An ester

19. What happens to a dilute solution of hydrochloric acid when water is added to it?

	pH	$H^+(aq)$ concentration
A	increases	increases
B	decreases	increases
C	decreases	decreases
D	increases	decreases

20. Which of the following solutions has the highest pH?

    A $0 \cdot 1$ mol l^{-1} ammonia solution

    B $0 \cdot 1$ mol l^{-1} sodium hydroxide

    C $0 \cdot 1$ mol l^{-1} ethanoic acid

    D $0 \cdot 1$ mol l^{-1} hydrochloric acid

21. $100 \, cm^3$ of a solution contains $0 \cdot 2$ moles of solute.

The concentration of the solution, in mol l^{-1}, is

    A $0 \cdot 002$

    B $0 \cdot 5$

    C   2

    D   5.

22. $2KOH(aq) + H_2SO_4(aq) \rightarrow K_2SO_4(aq) + 2H_2O(\ell)$

How many moles of potassium hydroxide are needed to neutralise $20 \, cm^3$ of sulphuric acid, concentration 1 mol l^{-1}?

    A $0 \cdot 01$

    B $0 \cdot 02$

    C $0 \cdot 03$

    D $0 \cdot 04$

Questions 23 and 24 refer to the following equation

$$Ba^{2+}(aq) + 2NO_3^-(aq) + 2Na^+(aq) + SO_4^{2-}(aq)$$

$$\downarrow$$

$$Ba^{2+}SO_4^{2-}(s) + 2Na^+(aq) + 2NO_3^-(aq)$$

23. The type of reaction represented by the equation above is

A   addition

B   displacement

C   neutralisation

D   precipitation.

24. The spectator ions present in the reaction above are

A $Na^+(aq)$ and $NO_3^-(aq)$

B $Na^+(aq)$ and $SO_4^{2-}(aq)$

C $Ba^{2+}(aq)$ and $NO_3^-(aq)$

D $Ba^{2+}(aq)$ and $SO_4^{2-}(aq)$.

25. In which of the following test tubes will a reaction occur?

A
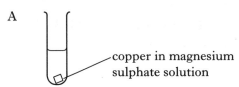
copper in magnesium sulphate solution

B
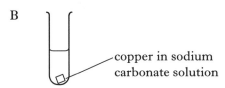
copper in sodium carbonate solution

C
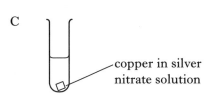
copper in silver nitrate solution

D
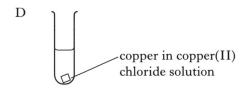
copper in copper(II) chloride solution

26. A cell produces a flow of electrons.

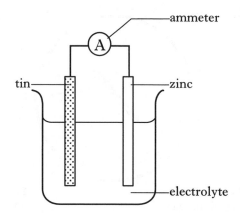

In the above cell, electrons flow from

A   zinc to tin through the ammeter

B   tin to zinc through the ammeter

C   zinc to tin through the electrolyte

D   tin to zinc through the electrolyte.

27. An acidic solution contains

A   only hydrogen ions

B   equal numbers of hydrogen and hydroxide ions

C   more hydrogen ions than hydroxide ions

D   more hydroxide ions than hydrogen ions.

28. Which of the following equations shows iron(II) ions being oxidised?

A $Fe^{3+}(aq) + e^- \rightarrow Fe^{2+}(aq)$

B $Fe^{2+}(aq) \rightarrow Fe^{3+}(aq) + e^-$

C $Fe(s) \rightarrow Fe^{2+}(aq) + 2e^-$

D $Fe^{2+}(aq) + 2e^- \rightarrow Fe(s)$

[Turn over

29.

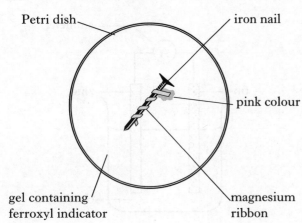

Petri dish

iron nail

pink colour

gel containing ferroxyl indicator

magnesium ribbon

Which ion gives a pink colour with ferroxyl indicator?

A $OH^-(aq)$

B $Fe^{2+}(aq)$

C $Fe^{3+}(aq)$

D $Mg^{2+}(aq)$

30. The coatings on four strips of iron were scratched to expose the iron. The strips were placed in salt solution.

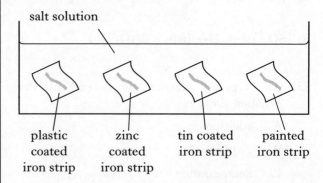

salt solution

plastic coated iron strip

zinc coated iron strip

tin coated iron strip

painted iron strip

The iron strip which would have rusted most quickly was the one which was

A plastic coated

B zinc coated

C tin coated

D painted.

Candidates are reminded that the answer sheet for Section A MUST be placed INSIDE the front cover of this answer book.

[Turn over for SECTION B on *Page ten*

Marks

SECTION B

50 marks are available in this section of the paper.

1. (*a*) The alkali metals, the halogens and the noble gases are the names of groups of elements in the Periodic Table.

Complete the table by circling a word in each box to give correct information about each group.

(Two pieces of correct information have already been circled.)

Group		
alkali metals	(metals) / non-metals	reactive / non-reactive
halogens	metals / non-metals	reactive / non-reactive
noble gases	metals / non-metals	reactive (non-reactive)

2

(*b*) Complete the table for the particle shown below.

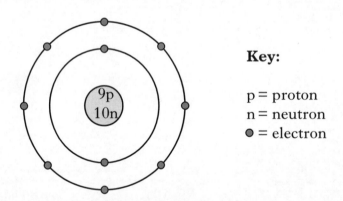

Key:

p = proton
n = neutron
● = electron

Atomic number	Symbol for the element	Mass number	Overall charge of the particle

2

(4)

Marks

2. (*a*) Sodium carbonate reacts with dilute hydrochloric acid.

    (i) Write the formula for sodium carbonate.

    _____

                                                                     1

    (ii) Name the salt formed when sodium carbonate reacts with hydrochloric acid.

    _____

                                                                      1

(*b*) A teacher used this reaction in an experiment to show that carbon dioxide puts out a flame.

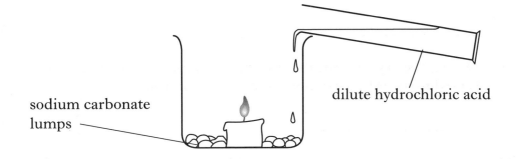

sodium carbonate lumps

dilute hydrochloric acid

Why would the candle burn for longer if an equal volume and concentration of ethanoic acid was used instead of the dilute hydrochloric acid?

_____

_____ 1

                                                                     (3)

[Turn over

DO NOT
WRITE I
THIS
MARGIN

Marks

3. (*a*) Zinc reacts with dilute hydrochloric acid producing hydrogen gas.

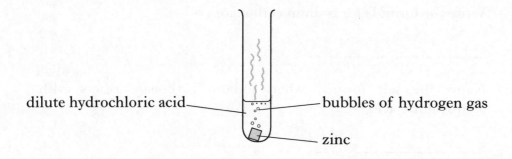

dilute hydrochloric acid — bubbles of hydrogen gas

zinc

(i) State the test for hydrogen gas.

_____

_____ 1

(ii) During the experiment, the test tube becomes warm. What term is used to describe a reaction which gives out heat?

_____ 1

(*b*) The rate of reaction between zinc and dilute hydrochloric acid can be followed by measuring the volume of gas given off during the reaction.

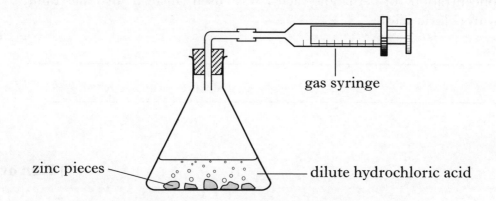

gas syringe

zinc pieces — dilute hydrochloric acid

Results	
Time (seconds)	**Volume of gas (cm³)**
0	0
10	20
20	40
30	58
40	72
50	80
60	

Marks

3. **(b) (continued)**

(i) Plot a line graph of the results of the reaction.

(Additional graph paper, if required, can be found on page 26.)

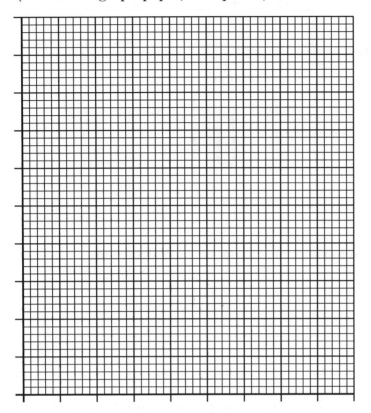

2

(ii) Predict the volume of gas which would have been given off after 60 seconds.

_____ cm^3 **1**

(c) Calculate the average rate at which gas is given off during the first 40 seconds of the reaction.

_____ cm^{3}s^{-1} **1**

(d) Why would increasing the concentration of the acid increase the rate of the reaction?

_____

_____   **1**

(7)

Page thirteen   **[Turn over**

Marks

4. In the **PPA "Electrolysis"**, copper(II) chloride is separated into its elements.

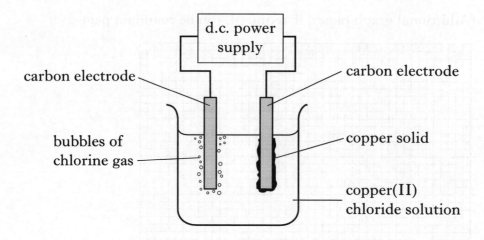

(a) Label the diagram to show the charge on each electrode.   **1**

(b) Describe how to smell chlorine gas safely.

_____

_____   **1**

(c) During the electrolysis reaction, copper ions are changed to copper atoms.

   (i) Why is this reaction described as reduction?

   _____   **1**

   (ii) At the end of this experiment 1·27 g of copper was deposited. Calculate the number of moles of copper deposited.

                                           _____ moles   **1**

                                                                            (4)

Marks

5. Plants release ethene gas. A build up of ethene in florists shops will cause flowers to wither quickly. Florists can use a solid titanium dioxide catalyst to break down ethene gas to make flowers last longer.

In the reaction, ethene reacts with the oxygen of the air to produce carbon dioxide and water.

(*a*) Balance the equation for this reaction.

$$C_2H_4(g) + O_2(g) \rightarrow CO_2(g) + H_2O(\ell)$$

1

(*b*) What type of catalyst is titanium dioxide in this reaction?

_____

1

(2)

[Turn over

Marks

6. Propene can take part in addition reactions.

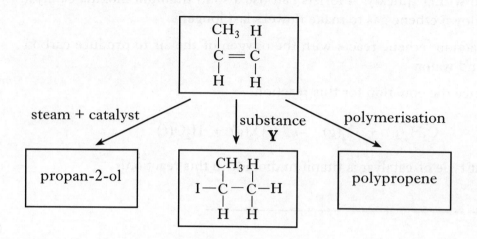

(a) (i) Draw the structural formula for propan-2-ol.

1

(ii) Give another name for the addition reaction that produces propan-2-ol.

_____

1

(b) Identify substance **Y**.

_____

1

(c) Draw a section of the polymer formed from propene showing three monomer units linked together.

1

(4)

Marks

7. In the **PPA "Hydrolysis of starch"**, dilute hydrochloric acid can be used to break down the starch.

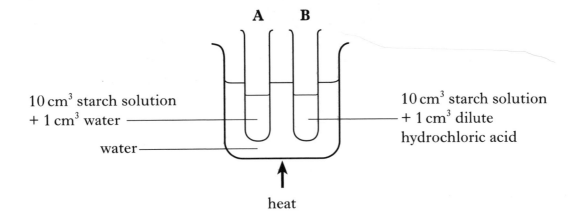

A   B

10 cm³ starch solution + 1 cm³ water

water

10 cm³ starch solution + 1 cm³ dilute hydrochloric acid

heat

(a) After heating with dilute hydrochloric acid, solid sodium hydrogencarbonate is added to each reaction mixture.

Why is sodium hydrogencarbonate added at this stage?

_____

_____ **1**

(b) Complete the table to show the results which should be obtained when the reaction mixtures are tested with Benedict's solution.

Reaction mixture	Observation on heating with Benedict's solution
A	
B	

1

(2)

[Turn over

DO NO
WRITE
THIS
MARGI

Marks

8. The diagram shows how an ester with a "pear drops" smell can be made.

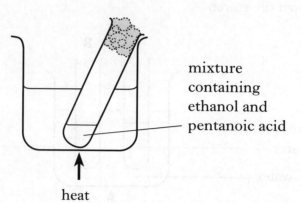

mixture
containing
ethanol and
pentanoic acid

heat

ethanol + pentanoic acid → ester + water

(a) (i) Name the ester formed.

_____ **1**

(ii) Why can this reaction be described as a condensation reaction?

_____

_____ **1**

(b) Polyesters are synthetic fibres.

Part of the structure of a polyester fibre is shown below.

$$-O-\overset{\overset{\displaystyle H}{|}}{\underset{\underset{\displaystyle H}{|}}{C}}-\overset{\overset{\displaystyle H}{|}}{\underset{\underset{\displaystyle H}{|}}{C}}-O-\overset{\overset{\displaystyle O}{\|}}{C}\!\!-\!\!\bigcirc\!\!-\!\!\overset{\overset{\displaystyle O}{\|}}{C}-O-\overset{\overset{\displaystyle H}{|}}{\underset{\underset{\displaystyle H}{|}}{C}}-\overset{\overset{\displaystyle H}{|}}{\underset{\underset{\displaystyle H}{|}}{C}}-O-\overset{\overset{\displaystyle O}{\|}}{C}\!\!-\!\!\bigcirc\!\!-\!\!\overset{\overset{\displaystyle O}{\|}}{C}-$$

(i) Why are polyester fibres described as synthetic?

_____ **1**

(ii) Circle an ester link in the polyester structure. **1**

(4)

Marks

9. Propane can be cracked to produce a mixture of smaller molecules.

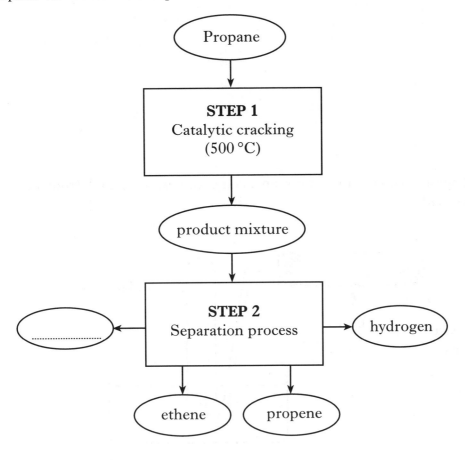

(a) Catalysts can be used to speed up a chemical reaction.

Give another advantage of using a catalyst.

_____

_____ 1

(b) Name the process used at step 2 to separate the product mixture.

_____ 1

(c) Complete the flowchart by naming the other product separated from
the mixture. 1

(3)

[Turn over

Marks

10. (*a*) Draw the full structural formula for propanoic acid.

1

(*b*) The diagram below shows how an alkane can be prepared from an alkanoic acid.

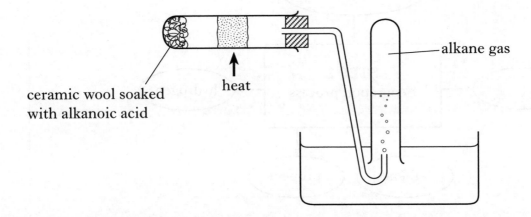

ceramic wool soaked
with alkanoic acid

heat

alkane gas

The equation for the reaction is:

alkanoic acid $\rightarrow$ alkane + carbon dioxide

Complete the table to show which alkanoic acid could be used to produce butane.

Alkanoic acid	Alkane
ethanoic acid	methane
propanoic acid	ethane
	butane

1

(2)

Marks

11. Fats are broken down in the body by hydrolysis.

$$
\begin{array}{c}
H \\
| \\
H-C-OH \\
| \\
fat \quad \rightarrow \quad H-C-OH \quad + \quad fatty\ acids \\
| \\
H-C-OH \\
| \\
H
\end{array}
$$

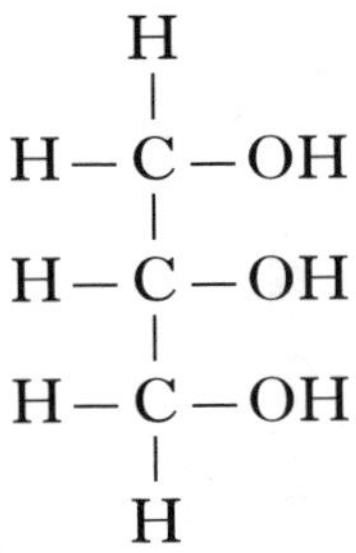

(a) When one mole of fat is hydrolysed, how many moles of fatty acids are produced?

_____ moles

1

(b) Name the molecule with the structure shown.

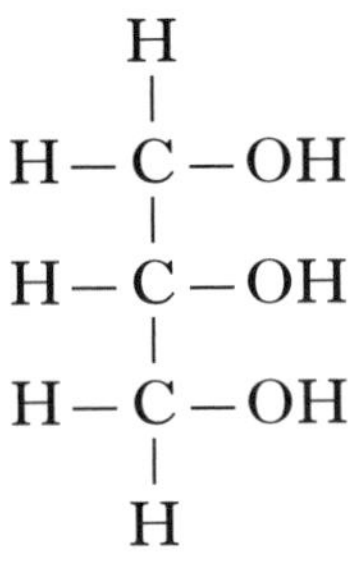

$$
\begin{array}{c}
H \\
| \\
H-C-OH \\
| \\
H-C-OH \\
| \\
H-C-OH \\
| \\
H
\end{array}
$$

_____

1

(c) Lipase is an enzyme which can catalyse the hydrolysis of fats in milk. Complete the diagram to show how the indicator colour would change after lipase was added to the test-tube.

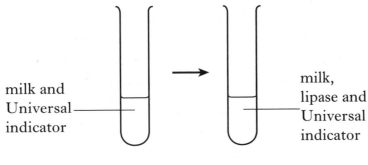

milk and
Universal
indicator

milk,
lipase and
Universal
indicator

Colour: **green**   Colour: _____

1

(3)

[Turn over

Marks

12. In the **PPA "Voltage"**, a student investigated the effect of changing the electrolyte.

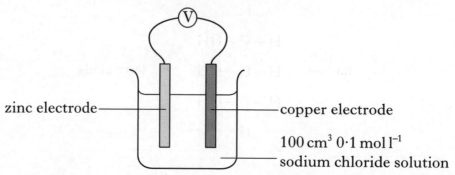

zinc electrode — — copper electrode

100 cm^3 0·1 mol l^{-1}
sodium chloride solution

(a) Draw and label the cell which would be used to measure the voltage produced when dilute hydrochloric acid is used as an electrolyte.

1

(b) For each cell two voltage readings were taken.

What should have been done before taking the second reading?

_____

_____

1

(c) The voltage of a cell is also affected by the metals used as electrodes. A student recorded the voltage and direction of flow for different cells.

Use the information in the table to predict the **direction of electron flow** and **voltage** you would expect when nickel and zinc electrodes are used.

Electrodes	Direction of electron flow	Voltage (V)
Cu/Zn	Zn → Cu	1·0
Cu/Ni	Ni → Cu	0·5
Fe/Ni	Fe → Ni	0·2
Fe/Zn	Zn → Fe	0·3
Ni/Zn		

2

(4)

Marks

13. The diagram represents the structure of a molecule of ammonia.

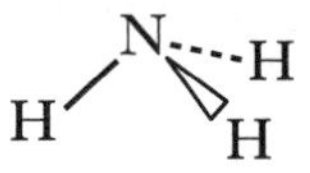

$$\underset{H}{\overset{N\cdots H}{\diagup}}\diagdown_{H}$$

(a) Why are the bonds between the nitrogen and the hydrogen in an ammonia molecule described as polar covalent?

_____

_____ **1**

(b) The equation shows what happens when ammonia gas dissolves in water.

$$NH_3(g) + H_2O(\ell) \rightleftharpoons NH_4^+(aq) + OH^-(aq)$$

   (i) Why is the solution formed alkaline?

   _____

   _____ **1**

   (ii) What does the sign, $\rightleftharpoons$, indicate about the reaction?

   _____

   _____ **1**

(c) Ammonia gas can be produced in the lab by heating ammonium chloride with sodium hydroxide.

Calculate the mass of ammonia produced by heating 10 g of ammonium chloride.

$$NH_4Cl + NaOH \rightarrow NaCl + H_2O + NH_3$$

_____ g **2**

(5)

Marks

14. The main source of magnesium in Britain is sea water.

(a) The first stage in the production of magnesium is to remove the magnesium ions from the sea water by reaction with calcium hydroxide solution.

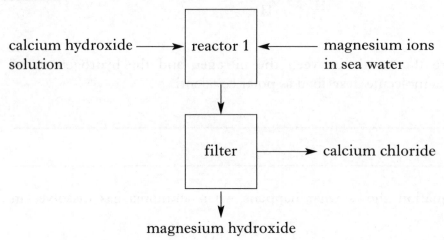

Why can magnesium hydroxide and calcium chloride be separated by filtration?

(You may wish to use page 5 of the data book to help you.)

_____

_____ 1

(b) In the next stage magnesium hydroxide is reacted with hydrochloric acid.

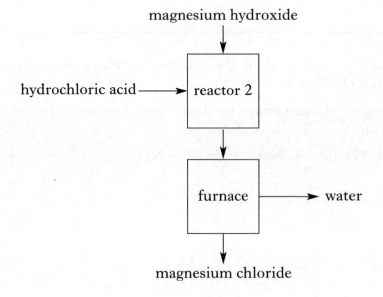

Name the type of chemical reaction occuring in reactor 2.

_____ 1

Marks

14. (continued)

(*c*) The last stage is the electrolysis of molten magnesium chloride.

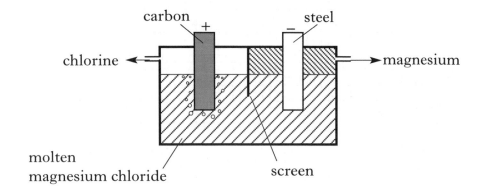

Write the ion-electron equation for the production of magnesium.

_____   **1**

(3)

[END OF QUESTION PAPER]

DO NO
WRITE
THIS
MARGI

ADDITIONAL SPACE FOR ANSWERS

ADDITIONAL GRAPH PAPER FOR QUESTION 3(*b*)(i)

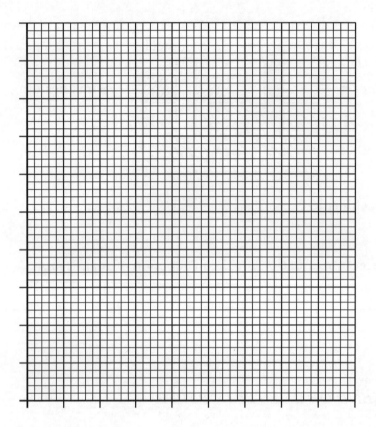

ADDITIONAL SPACE FOR ANSWERS

DO NOT
WRITE IN
THIS
MARGIN

ADDITIONAL SPACE FOR ANSWERS

[BLANK PAGE]

Chemistry
Intermediate 2
2004 (cont.)

14. (a)

Volume of water (cm³)	0	2	4	6

(b) When the solution in the flask turns blue-black
OR
When the tile is obscured
OR
When colour changes sharply
OR
When the cross is obscured

(c) The reaction rate is too slow
OR
Change too slow to distinguish end point

(d) $Na_2S_2O_8$ or $(Na^+)_2S_2O_8^{2-}$

15. (a) (i) Electrons flow onto the ship from the negative terminal of the power source

(ii) It contains ions/it is an electrolyte/it acts as an ion bridge

(b) Painting
OR
sacrificial protection
OR
attaching blocks of zinc/magnesium to the hull

16. (a) Universal or pH indicator or named pH indicator

(b) 0.123 mol⁻¹

Chemistry
Intermediate 2
2005

Section A

1.	B	11.	B	21.	C
2.	D	12.	B	22.	D
3.	A	13.	D	23.	D
4.	B	14.	B	24.	A
5.	D	15.	C	25.	C
6.	A	16.	B	26.	A
7.	A	17.	D	27.	C
8.	D	18.	C	28.	B
9.	A	19.	D	29.	A
10.	C	20.	B	30.	C

Section B

1. (a)

alkali metals	metals	reactive
halogens	non-metals	reactive
noble gases	non-metals	non-reactive

(b)

Atomic number	9
Symbol	F
Mass number	19
overall charge	−1

2. (a) (i) Na_2CO_3 or $(Na^+)_2CO_3^{2-}$
(ii) Sodium chloride/NaCl

(b) Rate of reaction is slower with ethanoic acid since it is a weak acid

3. (a) (i) Hydrogen burns with a "pop"
(ii) Exothermic

(b) (i)

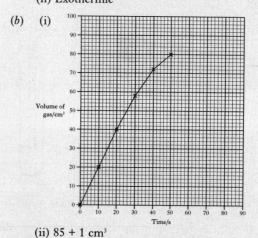

(ii) 85 + 1 cm³

(c) 1·8

(d) There would be a greater number of collisions therefore faster rate of reaction.

4. (a)

5. (a)

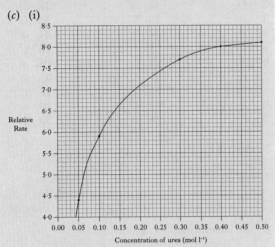

(b) 0·5 mole

(c) (i)

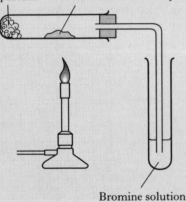

Relative Rate (y-axis)

Concentration of urea (mol l⁻¹) (x-axis)

(ii) 0·22 ± 0·01

6. (a) $2OH^-(aq) + 2H^+(aq) \rightarrow 2H_2O(l)$

OR

$OH^- + H^+ \rightarrow H_2O$

(b) It will turn it blue or blue/purple

(c) It can be used as a fertiliser/smelling salts

7. (a) Liquid paraffin   Aluminium oxide catalyst

Bromine solution

(b) The delivery tube should be removed from the bromine solution.

8. (a) Neutralisation

(b) 304 kg

9. (a) & (b)

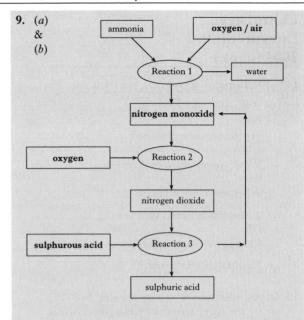

ammonia   **oxygen / air**

Reaction 1   →   water

nitrogen monoxide

oxygen → Reaction 2

nitrogen dioxide

sulphurous acid → Reaction 3 →

sulphuric acid

(c) It is completely dissociated into ions in water

10. (a) Amino acids

(b) (i)

(ii)

$$H-N-CH_2-C-OH$$

with H on N and O double bonded to C

(c) The papain is denatured/destroyed by high temperatures

11. (a) Condensation or esterification

(b) (i) Hydrogen or H_2

(ii) Sugar cane / sugar

12. (a) $2Cl^- + 2H_2O \rightarrow Cl_2 + 2OH^- + H_2$

(b) Hydroxide ions are repelled by the negatively charged groups on the PTFE membrane

(c) Thermoplastics

13. (a) Molecular / discrete

(b) Carbon, nitrogen and hydrogen

Chemistry
Intermediate 2
2003 (cont.)

14. (*a*)

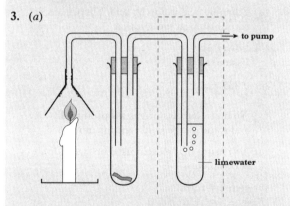

or isomer containing a $C\!=\!\!=\!\!C$ double bond

(*b*) Bromine solution

15. (*a*) Ethoxypropane

(*b*) Condensation

(*c*) 36°C

Chemistry
Intermediate 2
2004

SECTION A

1.	D	**16.**	B
2.	A	**17.**	B
3.	B	**18.**	D
4.	C	**19.**	C
5.	A	**20.**	A
6.	D	**21.**	D
7.	A	**22.**	C
8.	B	**23.**	B
9.	C	**24.**	A
10.	C	**25.**	D
11.	A	**26.**	A
12.	C	**27.**	D
13.	B	**28.**	B
14.	C	**29.**	D
15.	B	**30.**	A

SECTION B

1. (a) (i) $^{81}_{35}$ Br

(ii) 46 neutrons

(b) The percentages of each isotope are equal/the same/both 50%

2. (a) $2CO(g) + 2NO(g) \rightarrow 2CO_2(g) + N_2(g)$

(b) The catalyst is in a different state from the reactants.

3. (a)

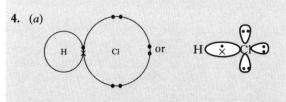

(b) Water / H_2O

(c) One of:
- Incomplete combustion/candle does not burn completely
- Not enough oxygen/air to burn completely

4. (a)

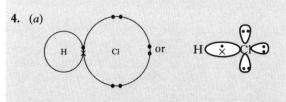

(b) Polar

4. (*a*) They provide energy

(*b*) Plants/vegetables

(*c*) Oils are less saturated (more unsaturated/have more double bonds) than fats
or
The forces between molecules are lower in oils

(*d*) By adding hydrogen across double bonds (by [catalytic] hydrogenation)/by making oils more saturated

5. (*a*) Going down a group the energy required to remove an electron decreases

(*b*) Going from sodium to argon there is a general increase in the energy required

(*c*) Between 400 and 750 kJ mol⁻¹

6. (*a*)

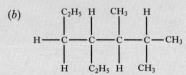

(*b*) The molecules have functional groups at each end of the molecule

7. (*a*) Methanol

(*b*) (i) 0.006 mol l⁻¹

(ii) 0.000015 (1.5×10^{-5})

8. (*a*) Ethanoic acid is a weak acid/There is an equilibrium in the solution

(*b*) Lower pH
Faster rate

(*c*) 0.835 mol l⁻¹

9. (*a*) No more bubbles of gas are produced and some of the solid remains unreacted

(*b*) To ensure that all acid used up/neutralisation is complete

(*c*) step 2 = filtration
step 3 = evaporation

(*d*) 24.1g

10. (*a*) Carbon dioxide and water (correct formulae accepted)

10. (*b*) (i)

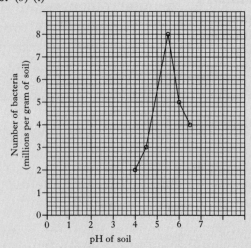

(ii) 5.5 or value given should coincide with candidate's graph

11. (*a*) 2,3-dimethylbutane

(*b*)

H——C——C——C——C—CH₃

with groups C₂H₅, H, CH₃, H above and H, C₂H₅, H, CH₃ below

12. (*a*) It has delocalised electrons/metallic bonding/electrons free to move

(*b*) Each atom gains an electron

(*c*)

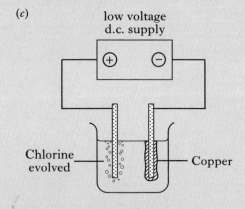

low voltage d.c. supply

Chlorine evolved — Copper

13. (*a*) $Zn(s) \longrightarrow Zn^{2+}(aq) + 2e^-$

(*b*) Tin less active than steel/iron (iron/steel more reactive)

or

Tin is lower down reactivity series than iron/steel (iron/steel higher up reactivity series)

or

Electrons flow from iron to tin iron/steel

(*c*) Cr_2O_3

Chemistry
Intermediate 2
2002 (cont.)

9. (b) (continued)

(ii) sulphur dioxide is a harmful (poisonous) gas

(c) (i) 50 seconds

(ii) As concentration increases there are a greater number of collisions and therefore a greater number of successful collisions resulting in an increase in the rate

10. (a) It is a very strong plastic or similar property

(b) They have two functional groups per molecule

(c)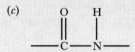

11. (a) The halogens

(b) (i) It indicates that the reaction is exothermic

(ii) An acid in which the molecules only partly dissociate to ions

12. (a) (i) The alkenes

(ii) Addition

(b) 604·7 g

(c) Greater degree of unsaturation in oils means more iodine will be able to react, therefore they will have a higher iodine number

(d) heterogeneous catalyst

13. (a) 31·9g

(b) (i) $Zn(s) + Cu^{2+}(aq) \longrightarrow Zn^{2+}(aq) + Cu(s)$

(ii) 0·0015

14. (a) As the length of the chain (or the mass) of the alkane increases the critical temperature increases

(b) higher than 187°C and lower than 234°C

Chemistry
Intermediate 2
2003

SECTION A

1.	B	16.	A
2.	C	17.	C
3.	B	18.	C
4.	A	19.	C
5.	C	20.	D
6.	D	21.	A
7.	B	22.	D
8.	B	23.	B
9.	D	24.	A
10.	A	25.	D
11.	D	26.	A
12.	B	27.	D
13.	C	28.	B
14.	A	29.	C
15.	D	30.	D

SECTION B

1. (a) A biological catalyst/protein

(b)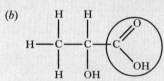

(c) The sugar might ferment producing alcohol/to prevent alcohol being produced

2. (a) A reaction which releases energy or heat or where products have less energy than reactants

(b) 2 moles

(c) Gas jar A contains more oxygen/(33% oxygen as opposed to 20% in air)

3. (a)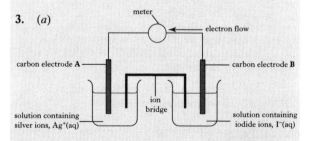

(b) $2Ag^+(aq) + 2I^-(aq) \longrightarrow 2Ag(s) + I_2(s)$

(c) To complete the circuit/to allow ions to flow/to act as a conductor

(d) Add starch, this gives a blue-black colour

Chemistry
Intermediate 2
2002

SECTION A

Part 1

1. A	14. A
2. D	15. C
3. B	16. D
4. B	17. B
5. C	18. A
6. D	19. D
7. C	20. A
8. C	21. A
9. C	22. B
10. B	23. D
11. C	24. A
12. B	25. A
13. D	

Part 2

26. (a) B

   (b) C and D

   (c) A and E

27. B and E

SECTION B

1. (a) $BaCl_2$; $Ba^{2+}(Cl^-)_2$

   (b) (i)

boiling point	Gas	1560°C
melting point	Liquid	963°C
	Solid	

   (ii) solid

2. (a) SiC

   (b) (i) It is covalent (has a covalent network structure) or all the outer electrons are used up in bonding

   (ii) A great deal of energy is required to break the strong covalent bonds

3. (a) $Fe \longrightarrow Fe^{2+} + 2e^-$

   (b) ferroxyl indicator

   (c)
   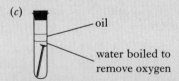
   oil
   water boiled to remove oxygen

4. (a) $C_4H_{10} + 13N_2O \longrightarrow 4CO_2 + 5H_2O + 13N_2$

   (b)
   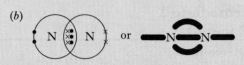
   or

5. (a) $D \longrightarrow C \longrightarrow A \longrightarrow B$

   (b) (i) Be careful when shaking the solution and wear gloves

   (ii)
   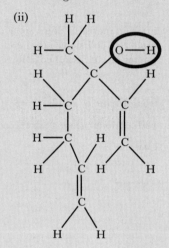

   (iii) esters

   (iv)
   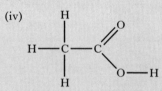

6. (a) Metal rods should be cleaned and dried

   (b) Any two from:
   keep electrolyte concentration same;
   keep electrolyte type same;
   temperature same;
   depth of rods in solution same;
   volume of electrolyte same

   (c) 0·7

7. (a) Acid rain, nitric acid

   (b) Neutralisation

   (c)
   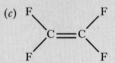

8. (a) Filtration

   (b) Carbon dioxide or ammonia (correct formulae acceptable)

   (c) Calcium chloride

   (d) 0·0448 mol l^{-1}

9. (a) $S_2O_3^{2-}(aq) + 2H^+(aq) \longrightarrow S(s) + SO_2(g) + 2H_2O(\ell)$

   (b) (i) by timing how long it takes for a cross marked on a piece of card under the beaker to be obscured

Chemistry
Intermediate 2
2001 (cont.)

9. (a) (i) relights a glowing splint

    (ii) % in air is too low/insufficient O_2 in air

    (b) 600·9g

10. (a) hydrolysis

    (b) proteins

    (c) activity increases

    (d) enzyme would be destroyed/enzyme denatured

    (e) sample would not turn blue/black when iodine added

11. (a) 0·00001

    (b) ethanoic acid is only partially dissociated into ions/is not all split up
    contains fewer H^+ (aq)

    (c) add Mg to HCl in a test tube

    rate measuring method, e.g. time for Mg to dissolve or observe vigour of effervescence

    Two factors affecting fairness, e.g. mass of Mg/vol or concentration of acid/temperature

    compare both reactions

12. (a) (i) $Ba^{2+}(aq) + SO_4^{2-}(aq) \longrightarrow Ba^{2+}SO_4^{2-}(s)$

    (ii) precipitation

    (b) 0·08 mol l^{-1}

13. (a) $Fe^{3+}(aq) + e^- \longrightarrow Fe^{2+}(aq)$

    (b) (i)

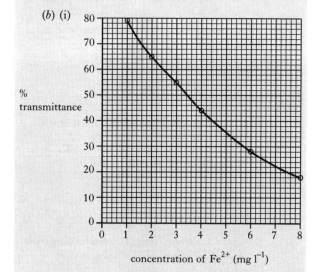

concentration of Fe^{2+} (mg l^{-1})

    (ii) 5 mg l^{-1} ($\pm$ 0·5 mg l^{-1} of value from graph)

14. (a) $2H_2 + O_2 \longrightarrow 2H_2O$

    (b) a liquid which contains ions

    (c) more molecules of gas collide with the catalyst surface

15. (a)

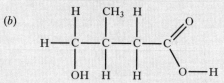

    (b)

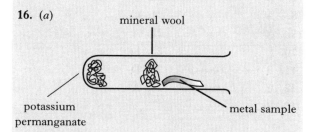

16. (a)

mineral wool

potassium permanganate

metal sample

    (b) test tube clamped horizontally/test tube not pointing at anybody/dry test tube/metal sample well separated from the $KMnO_4$/do not look directly at tube

© 2005 Scottish Qualifications Authority, All Rights Reserved
Published by Leckie & Leckie Ltd, 8 Whitehill Terrace, St Andrews, Scotland, KY16 8RN
tel: 01334 475656, fax: 01334 477392, enquiries@leckieandleckie.co.uk, www.leckieandleckie.co.uk

Chemistry
Intermediate 2 2001

SECTION A

Part 1

1.	D	14.	B
2.	A	15.	D
3.	A	16.	B
4.	C	17.	C
5.	B	18.	D
6.	B	19.	C
7.	D	20.	C
8.	A	21.	C
9.	A	22.	A
10.	B	23.	D
11.	B	24.	D
12.	C	25.	A
13.	B		

Part 2

26. C and D

27. C and E

28. (*a*) A and C

(*b*) D

SECTION B

1. (*a*)

Type of particle	Number of particles
proton	19
neutron	20
electron	19

(*b*) different numbers of neutrons/mass numbers

2. (*a*) carbon dioxide
    water

(*b*)

Position in series	Name	Molecular formula
1st	ethyne	C_2H_2
2nd	**propyne**	C_3H_4
3rd	butyne	**C_4H_6**

3. (*a*)

or $CH_2CHOCOCH_3$

(*b*) (i) its solubility in water

(ii)

or CH_3COOCH_3

4. (*a*) $Cu(NO_3)_2$

(*b*) $0·2 \times 0·25 = 0·05$

5. (*a*) heat the catalyst strongly
    flick the flame on to the liquid paraffin

(*b*) (i) unsaturated/contains $C{=}{=}{=}C$/contains alkenes

(ii) lift clamp stand/remove delivery tube from bromine before removing heat

(*c*) catalyst is in different state from reactants

6. (*a*) reduction/redox

(*b*) silver/mercury/gold

(*c*) (i) reaction with coke/carbon

(ii) $Fe_2O_3(s) + 3\ CO(g) \longrightarrow 2\ Fe(l) + 3\ CO_2(g)$

7. (*a*) allows the products to be identified/ions move in same direction

(*b*) the paper is bleached/turns white/loses colour

8. A (1) weigh dry evaporating basin

(2) place 10ml of $0·1$ mol l^{-1} NaCl solution in evaporating basin

(3) using tripod, gauze and bunsen burner, evaporate this to complete dryness

(4) weigh evaporating basin (which now has solid, dry NaCl in it)

(5) subtract reading from (1) from reading from (4) to give weight of NaCl that was in solution

B repeat (1) to (5) above for $0·2$ mol l^{-1} NaCl solution

weight from B should be twice weight from A

4. (b) Waft gas towards nose
or smell from a distance

(c) (i) copper ions gain electrons
(ii) 0·02 moles

5. (a) $C_2H_4 + 3O_2 \rightarrow 2CO_2 + 2H_2O$

(b) Heterogeneous

6. (a) (i)

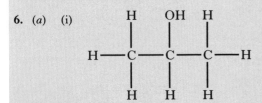

(ii) hydration

(b) Hydrogen iodide/HI

(c)

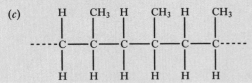

7. (a) To neutralise the excess acid

(b) A No colour change/stays blue
B Colour change to orange/brick red

8. (a) (i) Ethyl pentanoate
(ii) The molecules join together by
eliminating water.

(b) (i) They are man-made (made from oil)
(ii)

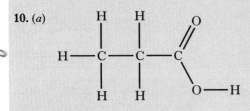

9. (a) Allows reactions to be carried out at lower
temperatures
or the catalyst can be re-used
or lowers activation energy

(b) Distillation

(c) Methane/CH_4

10. (a)

H   H   O
| | ⫽
H—C—C—C
| | \
H   H   O—H

(b) pentanoic acid

11. (a) 3 Moles

(b) Propan-1,2,3-triol or glycerol

(c) Red/orange (any acid colour with SQuA
indicator)

12. (a)

Zn   V   Cu
electrode   electrode

100 cm³ 0·1 mol l⁻¹
hydrochloric acid

(b) Electrodes removed, cleaned and dried,
replaced

(c) Ni/Zn   Zn → Ni   0·5V

13. (a) Because the electrons are not shared equally
or because nitrogen has a greater attraction for
electrons than hydrogen

(b) (i) Ammonia dissolves to produce
$OH^-(aq)$/hydroxide ions
(ii) The reaction is reversible

(c) 3·18 g

14. (a) Magnesium hydroxide is insoluble; calcium
chloride is soluble

(b) Neutralisation

(c) $Mg^{2+} + 2e^- \rightarrow Mg$